MW01614052

Marc Achtelig

**Technical Documentation Solutions Series:
Writing plain instructions**

*How to write user manuals, online help, and other forms of user
assistance that every user understands*

First Edition
English

indoition

indoition publishing e.K.
Goethestr. 24
90513 Zirndorf near Nürnberg, Germany

Tel.: *+49 (0)911/60046-659*
Fax: *+49 (0)911/60046-863*
Email: *info@indoition.com*
Internet: *www.indoition.com*

Author: Marc Achtelig
Editor: Andrea R. Winter
Proofreading: Elizabeth Meyer zu Heringdorf
Template design: Marc Achtelig
Cover design: Marc Achtelig
Cover picture: alexsl, iStockphoto
Printed by: Lightning Source

Trademarks
All terms mentioned in this book known by the publisher and by the author to be trademarks or service marks have been appropriately capitalized. However, the publisher and the author cannot attest to the accuracy of this information. Use of a term in this book should not be regarded as information whether a trademark or service mark does exist or does not exist.

Warnings and Disclaimer
Every possible effort has been made to ensure that the information contained in this book is accurate and complete, but the publisher and the author cannot accept responsibility for any errors or omissions, however caused. No warranty or fitness is implied. The publisher and the author shall have neither liability nor responsibility to any person or entity with respect to any loss or damage arising from the information contained in this book or from the lack of information not contained in this book. The book doesn't provide any individual advice; in particular, it doesn't provide any legal advice. Before shipping your products and before publishing any content, make sure that you follow all relevant standards, laws, and other regulations that are applicable for both your own country and for all countries to which you will sell and ship your products. All rules that are given in these standards, laws, and other regulations take precedence over the recommendations given in this book.

ISBN 978-3-943860-08-5

About the author

Marc Achtelig has been in the technical communication business since 1989.

After some years in the development of simulation and educational software, Marc joined one of the major German technical communication services providers as technical writer, information architect, and consultant. In 2004, he founded his own consulting business.

Marc was one of the pioneers and early evangelists of single source publishing, the approach to create printed manuals and online help systems from the same shared text base. He has published many articles and several books and has spoken at various national and international conferences.

He holds degrees in process engineering and industrial engineering.

For individual consulting services and trainings, contact Marc at *ma@indoition.com*.

Contents

1 How to use this book

Welcome to this book, your companion to writing clear user assistance.

What you will find in this book

Providing helpful user assistance isn't as simple as it may seem:

- If you've explained something thoroughly, this doesn't mean that your information is useful.
- If you've explained something accurately, this doesn't mean that you've explained it adequately.
- If your sentences are grammatically correct, this doesn't mean that readers understand what you want to say.

This book provides you with effective rules and tips that will help you to create content that is:

- well-structured
- clearly written
- easy to understand
- actually helpful

Each topic begins on a new page, so skimming through the book is easy. You don't have to read everything from start to finish. All topics are independent of each other. You don't have to read any particular topic to understand another one.

What you won't find in this book

The book provides clear rules and unambiguous recommendations. No boring theory, no musings, no highbrow grammar terms.

However, because each product and corporate style is different, there are often no ready-made, one-size-fits-all solutions. The book shows you what's important, it introduces you to the basic rules, and it can inspire you. However, the book can't make the final decisions for you.

Please consider the given rules as general recommendations, not as laws that must be followed slavishly.

ⓘ Important: The book can't provide individual advice; in particular, it can't provide any legal advice. Before shipping your products and before publishing any content, make sure that you also follow all relevant standards, laws, and other regulations that are applicable for both your own country and for all countries in which you sell your products. All rules that are given in these standards, laws, and other regulations take precedence over the recommendations given in this book.

This book is about perfection—but it isn't perfect

When reading this book, you might notice that we sometimes don't manage to follow our own advice. You might find typos, grammar mistakes, and things that could have been said more clearly. Ouch—sorry!

Believe us: We've tried hard to make everything perfect. We've used some of the best spelling checkers, grammar checkers, and writing enhancement software. We had the text double-checked by human editors. Yet, there are still some mistakes. Actually, no book is perfect.

What can you learn from this for your own documents?

- Don't be frustrated if you find errors in your own documents some time after the documents have been published. This is embarrassing, but it's normal.

- Don't think that you can eliminate mistakes completely. You can only minimize their number. Do so, but spend your resources wisely. Don't forget to optimize the contents, too.

- If you provide some real value to your readers, they will tolerate more errors than if you provide little value.

We hope that this book will provide enough value so that you will forgive us for the errors that we have made.

Have a good time reading and writing.

2 Writing

Everyone can write. However, not everyone can write so that everyone understands.

Characteristics of user-friendly style

Good user assistance is:

- correct and unambiguous
- written in a way that makes it easy to mentally process the given information
- written in a way that makes it easy to act upon the given instructions
- written in a way that makes it easy to remember the given information

Levels of writing involved

Creating clear, user-friendly documents involves all levels of writing. It starts with how you organize the information within a topic. It continues with how you structure paragraphs and build sentences. It ends with the choice of words. For details about each level, see the following sections:

- *Writing in general* 15
  Summarizes the key principles that you should follow on *all* levels of writing.
- *Writing topics* 45
  Tells you how to organize a topic's content, depending on the purpose of the topic.
- *Writing sections* 51
  Shows how you should organize the given information into paragraphs, how to add subheadings, and what to bear in mind when writing specific information types such as procedures or warnings.
- *Writing sentences* 87
  Shows how to build sentences that are grammatically simple, clear, and easy to understand.
- *Writing words* 113
  Shows how to choose words that add clarity rather than complexity.
- *FAQ: Spelling and punctuation* 161
  Provides a number of rules and working aids to help you with the most frequent spelling and punctuation issues.
- *FAQ: Grammar and word choice* 197
  Makes you aware of frequent grammar and word choice problems, such as the difference between the words *that* and *which*, or the correct use of the words *safety* and *security*.

- *FAQ: Standard terms and phrases* 26

  Often, you can have many names for one particular thing. To avoid confusion, you should always stick with one term. But which one is the right one? For example, should you call a program a *program*, **or an** *application*, **or** *software*? **Do you** *select* **an option, or do you** *choose* **it? This section gives recommendations on which terms to use.**

2.1 Writing in general

Write for the reader; don't write for your ego. Simpler is better.

Which style to follow?

Make reading your documents a positive experience. Write in a way that's:

- accurate and objective
- informative and helpful
- respectful and friendly
- positive and reassuring

General writing principles

The key overall principles when writing user assistance are:

- *Keep it simple and stupid* 16
- *Always start with the main point* 19
- *Talk to the reader* 21
- *Be specific* 23
- *Be concise* 26
- *Be consistent* 29
- *Be parallel* 32
- *Be positive* 35
- *Use the present tense* 38
- *Use the active voice* 39
- *Don't make judgments* 41
- *Don't say "please"* 43

Related rules

Writing topics 45
Writing sections 51
Writing sentences 87
Writing words 113

2.1.1 Keep it simple and stupid

Forget what you learned at school.

You're NOT writing an essay. You DON'T have to impress your teacher.

Provide information that EVERYBODY can understand—even readers who:

- don't speak the document's language as their first language
- aren't sitting in a silent office but who, for example, are standing in a noisy production hall
- don't have much time
- are frustrated because they didn't succeed without reading the manual

So **k**eep **i**t **s**imple and **s**tupid (KISS principle):

- Write short sentences.
- Use simple grammar.
- Use simple words.

Plain language is NOT evidence of poor education. Plain language is the foundation of clear user assistance.

✘ No: *If you want to exert influence on the contents of a document, access the submenu item **Edit** in the **File** menu after having opened the document file successfully.*

✔ Yes: *To edit a document:*

1. Open the document file.

*2. Choose **File** > **Edit**.*

✘ No: *Congratulations for buying this sophisticated, highly effective phone, which has been designed with your most vital communication needs in mind.*

✔ Yes: *You can use this phone to make phone calls.*

✔ Top: Leave out this sentence completely because it doesn't provide *any* useful information.

Can you measure simplicity?

Linguists have developed a number of indexes that attempt to measure the degree of complexity of a text. Some text editors have built-in functions to calculate these indices.

Don't take these indices too seriously. A comprehensibility index provides a rough estimate, but comprehensibility is determined by many more factors than average word length and average sentence length.

> **ⓘ Important:** Don't aim for a specific index value that's said to be "adequate" for the educational level of your audience. User instructions can't be too simple. Always aim for *maximum simplicity*. If you feel that your document might look too trivial for your audience, the document probably isn't too simple but too detailed. Try to identify things that you can omit.

Can you use software to guarantee simplicity?

On the market, there are a number of programs that can make suggestions about how to simplify a text. If you have access to one of these programs, go ahead and use it. Most of these programs can give you valuable feedback, however none of them can replace a human editor.

Also, don't forget that it's YOU who must structure and write clearly in the first place. If you don't, neither software nor a human editor will have much of a chance to improve your text.

Related rules

Be parallel 32

Be positive 35

Add syntactic cues 104

Make short sentences 88

Put the main thing into the main clause 91

Avoid parentheses and nested sentences 92

Feel free to start sentences simply 94

Feel free to end sentences simply 95

Use short, common words 114

Avoid abbreviations and acronyms 119

Use technical terms carefully 122

Always use the same terms 123

Avoid strings of nouns 128

Avoid stacks of modifiers 129

Avoid phrasal verbs 135

Avoid buzzwords 137

2.1.2 Always start with the main point

On all levels, provide the key message as soon as possible. Place the main points:

- in the first topic of a section
- on top of the page or screen
- at the beginning of a subsection rather than in the middle
- at the beginning of a paragraph rather than in the middle
- at the beginning of a sentence rather than in the middle
- in the first table column
- in the first table row
- on the left side of a figure (in languages that are read from left to right)

The position in front is the most prominent position:

- Readers assume that what comes right at the beginning is more important than what comes somewhere else.
- Readers remember better what comes at the beginning than what comes somewhere else.
- When readers skim a text, what comes at the beginning is easier for them to find than what comes somewhere else.
- Readers who don't read the full topic at least read the key message at the beginning.

How to identify the main point

It's your job as the author to decide what's most important and thus what becomes the main point. Often, the main point is:

- what most users need to know
- what users need to know early
- what's not optional
- what may cause an error, damage, injury, or death
- what's a prerequisite for an action
- what users must find or do first

For examples and for additional criteria, see *Watch the order of words* 96.

Related rules

Watch the order of words 96

2.1.3 Talk to the reader

1 Talk to the reader directly ("You can ...."). Talking directly to the reader increases attention and avoids ambiguity (see also *Use the active voice* 39).

Don't use the passive voice ("... can be done.").

Don't talk about the user ("Users can ...."). Don't use phrases with "one" ("One can ....").

Don't be afraid of giving commands. It's your job to tell your readers clearly what to do.

Write as though you were talking to your readers in a friendly, straightforward way (conversational style). Keep it simple, make short sentences, and use the same short, everyday words that you use when talking to co-workers.

When giving recommendations, it's acceptable to use *we*. Don't use the passive voice to avoid the editorial *we*. Often, however, the best solution is to create a sentence with *you*.

Exceptions:

2 If you're writing for developers or for administrators, use second person to refer to your reader (the developer or administrator), but use third person to refer to the reader's end user.

3 In error messages and troubleshooting information, it can be more polite to use a passive construction rather than to tell users right away that a problem is their own fault (see *Errors* 326).

4 In tutorials, a passive construction is sometimes appropriate to distinguish general information from a prompt to act.

1

✘ **No:** *The button must be pressed.*

✘ **No:** *The button must be pressed by the user.*

✘ **No:** *Users must press the button.*

✘ **No:** *One must press the button.*

✔ **Yes:** **Press the button.**

✘ **No:** *The Print dialog provides the possibility to change the printer settings.*

✔ **Yes:** **In the Print dialog, you can change the printer settings.**

✘ **No:** *In this section, the installation of the program will be shown.*

✔ **Yes:**   *In this section, we show you how to install the program.*
✔ **Top:**   *In this section, you'll learn how to install the program.*

✘ **No:**   *It's recommended to use a shielded cable.*
✔ **Yes:**   *We recommend using a shielded cable.*
✔ **Top:**   *For best performance, use a shielded cable.*

2

✘ **No:**   *Administrators can reset passwords so that users are able to create a new password.*
✘ **No:**   *Passwords can be reset so that users are able to create a new password.*
✔ **Yes:**   *You can reset passwords so that users are able to create a new password.*

3

✘ **No:**   *You've made a serious mistake. Next time, read the manual before you try this.*
✔ **Yes:**   *A mistake has been made. More information can be found in the manual.*

4

✘ **No:**   *To make a phone call, type the telephone number. Now you try it: Type a friend's telephone number.*
✔ **Yes:**   *Phone calls are made by typing the telephone number. Now you try it: Type a friend's telephone number.*

Related rules

Writing procedures 56
Use the active voice 39
Use strong verbs 132
Don't say "please" 43

2.1.4 Be specific

When reading instructions, users are looking for clear answers.

- Don't be vague.
- Don't be ambiguous.

You know the product that you're describing—readers don't. What's clear to *you* may not be clear at all to *your audience*. Also, bear in mind that most readers don't read manuals from start to finish, so they only see a small part of the whole.

Unclear or ambiguous text has a serious impact on the perceived quality of your document:

- If readers *do* notice that a phrase is unclear or ambiguous, this results in uncertainty. Ambiguous texts don't inspire confidence.
- If readers *don't* notice that a phrase is unclear or ambiguous (which often happens), this may result in misunderstanding and failure.
- If an unclear or ambiguous phrase goes unnoticed by translators, translated versions of your document may be plain wrong. In this case, *all* readers who read the translated version get the wrong information.

The key rule on how to be specific is to avoid all sorts of vague terms.

Note:
Being specific is more important than being concise (see *Be concise* 26).
Don't write ambiguous sentences because you want to make them as short as possible. If necessary, don't hesitate to repeat a word, or add a syntactic cue (see *Feel free to repeat a word* 10 and *Add syntactic cues* 104).

✘ **No:** *If you've filled in all fields correctly, the results window should appear.*

✔ **Yes:** *If you've filled in all fields correctly, the results window appears.*

(If you doubt that your product works as intended, don't show your doubt to the reader. Always describe the intended behavior or usual condition.)

✘ **No:** *The action should be finished quickly.*

✔ **Yes:** *You need to finish the action within one minute.*

✘ **No:** *The program can also import a number of other formats.*

✗ No: The program can also import Word files, etc.

✔ Yes: **The program can also import Word files, PDF files, and XML files.**

✗ No: If necessary, turn the lights on.

✔ Yes: **If it's dark and the lights are off, turn them on.**

✗ No: between 7 and 11

(Unclear: Are 7 and 11 included?)

✔ Yes: **from 7 through 11**

✗ No: The file set includes common files that are used in many web applications.

✔ Yes: **The file set includes files that are shared between web applications on the web server.**

or:

The file set includes files that many web applications typically use.

✗ No: We will successfully install your washing machine.

✔ Yes: **We will unpack your washing machine, set it up, connect it, and get it working.**

✗ No: Press any key to continue.

✔ Yes: **Press [Enter] to continue.**

(Note: Name a specific key even if other keys will do the same job. For users, it's faster to look for the [Enter] key than having to choose a key by themselves. In addition, it prevents any feeling of uncertainty.)

Typical examples of vague terms

- and so on
- and/or
- can
- corresponding
- etc.
- may
- maybe
- object

- *ought to*
- *quite*
- *rather*
- *respectively*
- *should*
- *some*

Related rules

2.1.5 Be concise

Omit all words and syllables that are nothing but empty calories.

Every word and character saved is a step toward more clarity. The only exception to this rule is: Don't be concise at the expense of clarity. If you need more words to be more specific or to avoid ambiguity, go ahead and include them (see *Add syntactic cues* [104], *Be clear about what you're referring to* [106], and *Feel free to repeat a word* [107]).

The key to avoiding empty calories in your documents is to be aware why you may be tempted to add them:

- When you aren't sure about the facts that you describe, you might be tempted to conceal your uncertainty by adding something vague.
- You might be tempted to impress your readers with your sophisticated language skills or with your profound domain knowledge.
- You might be tempted to impress your boss with the number of pages that you've produced.
- You might be tempted to impress customers with the number of pages because you think that a big manual makes your product look like it's worth the money.
- You just don't care and write down something quickly because you don't like writing manuals and want to complete this task as soon as possible.

Resist these temptations.

✘ No: *The program can handle the following four file formats: A, B, C, and D.*

✔ Yes: **The program can handle the file formats A, B, C, and D.**

(In this sentence, the relevant fact is which file formats the program supports. The number of formats ("four") is irrelevant, so leave it out. Also you can leave out the phrase "the following" without any loss of information.)

✘ No: *The new car is faster and will break down less often.*

✔ Yes: **The new car is faster and more reliable.**

✘ No: *The cable is about 10 meters in length.*

✔ Yes: **The cable is 10 meters long.**

✘ No: *It has a rectangular shape.*

✔ Yes: *It's rectangular.*

✘ No: *You should have some experience within a Unix environment.*
✔ Yes: *You need to have some Unix experience.*

✘ No: *If you're a user who has experience in this field, use expert mode.*
✔ Yes: *If you're an experienced user, use expert mode.*

✘ No: *The program isn't able to print.*
✔ Yes: *The program can't print.*

✘ No: *It's necessary to enter a value.*
✘ No: *You're required to enter a value.*
✔ Yes: *You must enter a value.*

✘ No: *You can format the table by means of the toolbar.*
✔ Yes: *To format the table, use the toolbar.*

✘ No: *In order to print the file, choose the menu command File > Print.*
✔ Yes: *To print the file, choose **File** > **Print**.*

✘ No: *It takes a longer period of time to write a user manual than to read it.*
✔ Yes: *It takes longer to write a user manual than to read it.*

Related rules

Talk to the reader `21`
Use the active voice `39`
Don't say "please" `43`
Make short sentences `88`
Avoid parentheses and nested sentences `92`
Feel free to start sentences simply `94`
Feel free to end sentences simply `95`
Use short, common words `114`
Watch for "...ed" `116`
Watch for "the ... of" and for "of the" `117`
Watch for opening "It ..." and "There ..." `118`
Use contractions `125`

2.1.6 Be consistent

A consistent document is easy to read. The reader can fully focus on the content.

In addition, a consistent document makes a professional impression, which increases your credibility and builds up confidence.

Document consistency is a result of:

- consistent design and formatting
- parallel structures and phrases (see *Be parallel* 32)
- consistent spelling, punctuation, and choice of words (see *Always use the same terms)* 123

Aim for consistency not only within each document but also among all documents.

✖ **No:**   Documentation for product A:
- Setting up Product A
- First Steps with Product A
- Using Product A
- Product A Technical Reference

Documentation for product B:
- Installing Product B
- Getting Started with Product B
- Product B User's Guide
- Product B Developer's Guide

✔ **Yes:**   *Documentation for product A:*
- *Product A Installation Guide*
- *Product A Getting Started Guide*
- *Product A User's Guide*
- *Product A Developer's Guide*

Documentation for product B:
- *Product B Installation Guide*
- *Product B Getting Started Guide*

- *Product B User's Guide*
- *Product B Developer's Guide*

Tips for obtaining consistency

Consistency can be hard to achieve, especially when more than one author works on the same document. But even if you're the only author, it's often difficult to remain consistent in the long run.

To remain consistent:

- Create common document templates that all authors are obliged to use.
- Keep track of your preferences in a terminology and preferences list.

Sample structure of a simple terminology list

In many cases, you don't have to use a dedicated terminology database or terminology management system to achieve a consistent use of terms. Even a simple, short terminology list written in any spreadsheet program can work wonders.

For example, a basic terminology list could look like this:

Term to use	Terms NOT to use	Comments
computer	client device machine PC unit workstation	When referring to computer networks, it's OK to use *client* in this particular context.
...	...	...
...	...	...

Don't use the terms listed in the "Terms NOT to use" in your visible texts, but do add them as index keywords to support readers who use them.

Tip:
Add all terms that you *don't* want to use to the exclusions list of your spelling checker so that the spelling checker marks them as *wrong*. This is an inexpensive and effective way to identify undesirable terms automatically, especially if your spelling checker checks the spelling as you type (live spelling check).

Related rules

Always use the same terms 123

FAQ: Standard terms and phrases 26

Be parallel 32

2.1.7 Be parallel

In technical writing, repetitive structures aren't a sign of weak style but enhance readability.

Parallel structures make the content more predictable. Readers don't have to mentally process a new structure but can, instead, attend to the words alone.

When possible, structures should be parallel:

- between sentences
- within a sentence

Parallelism is especially important in phrases with *and* and *or*.

Parallelism in headings

Try to keep all headings within a chapter, section, or other unit grammatically parallel, especially those on the same level. If it doesn't make sense to keep all headings parallel, try to keep at least as many subsequent headings as parallel as possible.

✖ **No:** *Washing of trucks*
Washing cars
How can you wash a motorcycle?
How to wash a bicycle

✖ **No:** *Washing trucks*
Washing cars
Washing a motorcycle
Washing a bicycle

✖ **No:** *Washing trucks*
Cleaning cars
Giving a wash to motorcycles
Cleansing bicycles

✔ **Yes:** **Washing trucks**
Washing cars
Washing motorcycles
Washing bicycles

Parallelism in lists and tables

Don't mix full sentences with sentence fragments (see also *Writing lists* 64 and *Writing tables* 68).

✘ No: In a document, white space is important for the following reasons:

- visual separation of sections
- attention focus
- white space breaks content into smaller chunks

✔ Yes: In a document, white space is important because it:

- visually separates sections
- focuses attention
- breaks content into smaller chunks

Parallelism in procedures

Typically, begin all steps with a verb (see also *Writing procedures* 56).

✘ No: To print a picture:

1. Open the image file.
2. Choose File > Print > Options.
3. Next, select the option Photo Quality.
4. Finally, the Print button must be clicked.

✔ Yes: To print a picture:

1. Open the image file.
2. Choose the menu command **File** > **Print** > **Options**.
3. Select the option **Photo Quality**.
4. Click the **Print** button.

Parallelism within sentences

Balance parts of a sentence with their correlating parts (nouns with nouns, prepositional phrases with prepositional phrases, and so on).

Note:
Sometimes you need to repeat a word. This is perfectly OK (see *Feel free to repeat a word* 101).

✘ No: The program can be used to manage annual reports, budgets, and financial planning.

✔ Yes: You can use the program for managing annual reports, for budgeting, and for financial planning.

✔ Yes: You can use the program to manage annual reports, budgets, and financial plans.

✘ No: *Whether at home or when working, DemoProduct helps you.*

✔ Yes: *Whether at home or at work, DemoProduct helps you.*

✔ Yes: *When at home or when working, DemoProduct helps you.*

✘ No: *Your alternatives are to use feature A or using feature B.*

✔ Yes: *Your alternatives are using feature A or using feature B.*

✔ Yes: *Your alternatives are to use feature A or to use feature B.*

✘ No: *The product can be used by children and adults.*

✔ Yes: *The product can be used by children and by adults.*

✔ Top: *Children and adults can use the product alike.*

(This version is better because it avoids the passive voice.)

✘ No: *The traffic light may be red or green. Green means that you can go, red means that you must stop.*

✔ Yes: *The traffic light may be red or green. Red means that you must stop, green means that you can go.*

(Here, the second sentence uses the same order of colors as the first sentence: first red, and then green.)

Related rules

▶ *Be consistent* 29

Feel free to repeat a word 10

2.1.8   Be positive

1 State your points positively when you can.

- Positive sentences convey a better image of your product.
- Positive sentences are more engaging.
- Positive sentences reinforce the readers' memory.

It's not always possible to make your writing positive, but in most cases you can swap the "don't" action with the "do" action.

2 Don't force yourself to be positive in warning messages. Sometimes you can, but often you *must* use a negative sentence here when something might cause serious damage, injury, or death.

3 You also sometimes can't avoid negative language when you need to say what the product can't do.

4 Avoid double negations. They are often very hard to understand and a frequent source of confusion.

1

✖ No: Don't put statements in the negative form.

✔ Yes: Put statements in the positive form.

✖ No: Don't switch off the computer until you've saved your documents.

✔ Yes: Save your documents before you switch off the computer.

2

✖ No: Don't apply if the temperature is below 10 degrees Celsius.

✔ Yes: Only apply if the temperature is at least 10 degrees Celsius.

(Here it's possible to use positive language.)

✖ No: Keep the hair dryer out of water.

✔ Yes: Don't immerse the hair dryer in water.

(In this case, a positive statement isn't strong enough.)

✖ No: You can't store more than 1,000 phone numbers.

✔ **Yes:**   *You can store up to 1,000 phone numbers.*

(Here it's possible to use positive language.)

3

✖ **No:**   *The translation program can handle most European languages.*

✔ **Yes:**   **The translation program can handle all European languages except Finnish, Slovakian, and Albanian.**

4

✖ **No:**   *It's not unlikely that ....*

✔ **Yes:**   *It's likely that ....*

✖ **No:**   *Make sure that you don't use unlicensed software.*

✔ **Yes:**   *Make sure that you only use licensed software.*

Examples of negative words

Note that a sentence doesn't have to include the word *not* to be negative. Many words are negative on their own, such as:

- *bad*
- *don't*
- *error*
- *hardly*
- *lack of*
- *less*
- *missing*
- *never*
- *no, none, nobody, nowhere*
- *problem*
- *without*

Prefixes also often indicate the negative nature of a word. Typical examples are:

- *a...*
- *de...*
- *dis...*

- *il...*
- *im...*
- *in...*
- *ir...*
- *non...*
- *un...*

2.1.9   Use the present tense

> **1** Write as if you were talking over the reader's shoulder. Write as if the reader was using the application right now. Use the present tense.
>
> Using the present tense implies that things always happen as described. This inspires confidence.
>
> **2** Save the future tense for things that will happen in the future.
>
> The future tense is also appropriate if the product or a particular feature that you describe isn't yet available.

1

✘ **No:** Click the Send button. Your email will be sent to the recipient.

✔ **Yes:** **Click the Send button. The program now sends your email to the recipient.**

✘ **No:** The Printer Options window will appear.

✔ **Yes:** **The Printer Options window appears.**

2

✘ **No:** You receive an answer to your email within one hour.

✔ **Yes:** **You will receive an answer to your email within one hour.**

✘ **No:** The upcoming version is able to ....

✔ **Yes:** **The upcoming version will be able to ....**

2.1.10   Use the active voice

1 Use the active voice rather than the passive voice.

- The active voice always makes clear whether the reader must act or whether the system acts automatically.

- The active voice is shorter and easier to understand than the passive voice. It directly tells the users what to do. This is particularly important in warning messages.

Note:
The rule to avoid the passive voice sometimes conflicts with the rule to put the more important information before the less important information within a sentence. If this happens, use the active voice anyway or rephrase the sentence.

Exceptions:

2 You can use the passive voice intentionally if you want to avoid pointing a finger at someone. In error messages and troubleshooting information, using the passive voice can be a polite way to tell users that a problem is their own fault (see *Errors* 326).

3 In tutorials, a passive construction is sometimes appropriate to distinguish general information from a prompt to act.

1

✗ No: *The active voice is to be used.*

✔ Yes: *Use the active voice.*

✗ No: *To enter the room, the door must be unlocked.*

✗ No: *The door must be unlocked to enter the room.*

✗ No: *If users want to enter the room, they must unlock the door.*

✗ No: *One must unlock the door before the room can be entered.*

✗ No: *In order to enter the room, it's necessary that the door gets unlocked.*

✔ Yes: *To enter the room, unlock the door.*

✗ No: *It is recommended to ....*

✔ Yes: *We recommend ....*

✗ No: *Now a backup can be made.*

✔ Yes: *Now you can make a backup.*

or:

Now the system automatically makes a backup.

✘ No: *More information can be found in the appendix.*

✔ Yes: *You can find more information in the appendix.*

2

✘ No: *You've made a serious mistake. Next time, read the manual before you try this.*

✔ Yes: *A mistake has been made. More information can be found in the manual.*

3

✘ No: *To make a phone call, type the telephone number. Now you try it: Type a friend's telephone number.*

✔ Yes: *Phone calls are made by typing the telephone number. Now you try it: Type a friend's telephone number.*

Related rules

Writing procedures 56

Talk to the reader 21

Use strong verbs 132

Watch for "...ed" 116

2.1.11 Don't make judgments

Don't ruin your credibility by making judgments and by talking between the lines.

1 Get rid of adjectives and unnecessary filler words.

- Remember that you're writing user assistance, not sales materials.

- In particular, avoid judgment calls on ease and simplicity. For example, avoid words such as *easy*, *simple*, *user-friendly*, *just*, *very*, etc. For your readers, your product isn't that simple—if it were that simple, they wouldn't read your instructions. By claiming that it's easy to use, you offend them.

2 Avoid superlatives.

3 Use adjectives and superlatives only where they distinguish different objects.

1

✖ **No:** *Just fill in this simple form.*

✔ **Yes:** *Fill in the form.*

✖ **No:** *DemoSoft also makes writing letters extremely easy. You only need to enter your text into the user-friendly interface. and DemoSoft takes care of all the annoying formatting tasks.*

✔ **Yes:** *You can also use DemoSoft for writing letters. You only need to enter your text. DemoSoft takes care of the formatting.*

✖ **No:** *Of course, you must plug in the cable before ....*

(The phrase *of course* might be mistaken for arrogance.)

✔ **Yes:** *Don't forget to plug in the cable before ....*

(Implies that the reader might be forgetful—so only use this version if forgetting to plug in the cable is a very common mistake.)

✔ **Top:** *Plug in the cable before ....*

2

✖ **No:** *Use this innovative, environmentally friendly feature to print multiple pictures onto one single sheet of paper.*

✔ **Yes:** *Use this feature to print multiple pictures onto one page, so you can save paper.*

3

✔ **Yes:**   *Press the red button.*

✔ **Yes:**   *Remove the longest item.*

Related rules

Avoid unnecessary qualification 130

Avoid buzzwords 137

2.1.12 Don't say "please"

1 When you're giving instructions, you're *not* asking readers for a favor.

For this reason, using the word *please* is inappropriate and just bloats your text. (Also, where would you stop? Would you say *please* in every sentence? Once in a paragraph? Once in a topic? It just doesn't make sense.)

Using the word *please* can even be misleading because it implies that an action is optional, which in many cases probably isn't the case.

2 The only time when it's appropriate to say *please* is when you're apologizing for a problem, or when you're actually asking for a favor.

1

✘ No: *Please insert the program CD into the disc drive.*
✔ Yes: *Insert the program CD into the disc drive.*

2

✔ Yes: *If you get an error message, please contact support.*
✔ Yes: *Can we improve this manual? Please send your feedback to feedback@yourdomain.com.*

Related rules

Talk to the reader 2↑

2.2 Writing topics

Don't mix information types. For each topic, decide on one specific information type, and then write the text specifically for this information type.

Choose a topic title (heading) that clearly communicates both the topic's information type and the topic's content.

When you write the topic's body text, don't assume that readers have read the heading. This often isn't the case, particularly in online help when users come to a topic via search or via a link.

✘ No: heading: *Printing Reports*
first sentence in topic: *This can be done with the Print command.*

✔ Yes: heading: *Printing Reports*
first sentence in topic: *You can print Reports with the Print command.*

Rules for standard topic types

Follow the specific writing rules for the topic's information type:

- *Writing "Concept" topics* 46
- *Writing "Task" topics* 47
- *Writing "Reference" topics* 49

Related rules

Writing in general 15

Writing sections 51

Writing sentences 87

Writing words 113

2.2.1 Writing "Concept" topics

"Concept" topics provide basic information and describe how things are related. They contain overviews, definitions, rules, and guidelines.

The goal of a Concept topic is to help users understand the basic principles so that they can make informed decisions and explore the product largely on their own.

Concept topics are primarily for beginners.

Dos and don'ts

- In Concept topics, don't swamp users with details. Mention only the basics that users should actually memorize.

- Instead of just mentioning facts, provide explanations. Show how things are related.

- Use pictures if you can.

- Use examples and metaphors that help readers relate the information to their own work, knowledge, and experience.

- Don't give instructions. For this purpose, use a Task topic instead (see *Writing "Task" topics* 47).

- Don't provide tables, lists, or diagrams that are mainly intended for users to look up specific data. For this purpose, use a Reference topic instead (see *Writing "Reference" topics* 49).

Related rules

Writing "Task" topics 47

Writing "Reference" topics 49

2.2.2   Writing "Task" topics

"Task" topics describe how to accomplish a specific job. The goal is to enable users to accomplish the job without reading a large amount of text.

Typically, Task topics consist of:

- Optional: A brief description why the task needs to be performed.
- Optional: Prerequisites that must be fulfilled before the task can be performed.
- Required: A series of steps that users can follow to produce the intended outcome (see *Writing procedures* 56).
- Required: Warnings if the procedure involves any dangers (see *Writing warnings* 72).
- Required: A brief description or picture of the procedure's result so that readers can verify that they've followed all steps correctly.
- Optional: Links to related Concept topics and Reference topics.
- Optional: Link to a Task topic that describes the subsequent task within a larger context or workflow.

Dos and don'ts

- If the task is so complex that you can't describe it within one topic, just outline the major steps and link to other topics that each contain the description of one subtask.
- Don't explain *why* things are done in a particular way.

  If you need to provide extensive background information, link to a Concept topic (see *Writing "Concept" topics* 46). However, it's OK to include a short conceptual element if this explanation helps users to avoid mistakes or if it helps users to better remember the steps of the procedure.

- If the procedure involves any hazards, add a warning immediately before the step that involves the hazard (see *Writing warnings* 72).
- It's not the job of a Task topic to provide technical data or detailed reference information.

  If required, link to a Reference topic (see *Writing "Reference" topics* 49). However, it's OK to give a short list of parameters or brief explanations of options if setting these parameters must be done by all users who perform the procedure. To also make this information accessible to advanced users who don't read the Task topic, add the same information to a Reference topic as well—even if this means redundant information.

- Describe only one way to accomplish a goal. Choose the most common and easiest way. When documenting software, describe keyboard shortcuts in a separate reference of shortcuts.

How many procedures should go into one Task topic?

According to the principles of structuring content, each topic should be self-contained.

Depending on how you write the heading, you can design a Task topic to contain a family of procedures or a single procedure.

The advantages of single-procedure Task topics are:

- The topics are shorter.
- Users can access each procedure directly.

The advantages of multiple-procedure Task topics are:

- Procedures stand in a broader context, which often helps to explain their use and relationships.
- Related procedures are obvious and don't have to be linked to.
- Clustering closely related procedures reduces the total number of topics.
- Having fewer topics makes navigation easier (readers must scan fewer entries in the table of contents; there are fewer links to related topics).

Example of a single-procedure Task:

topic title: *Deleting Files*

Example of a multiple-procedure Task:

topic title: *Handling Files*
first subsection: *Creating Files*
second subsection: *Editing Files*
third subsection: *Deleting Files*

Related rules

Writing "Concept" topics 46

Writing "Reference" topics 49

Writing procedures 56

2.2.3 Writing "Reference" topics

"Reference" topics contain detailed factual material, often in the form of tables and diagrams.

The goal of a Reference topic is to provide specific information (often data) quickly and selectively. The information is given in full detail.

Reference topics are primarily for intermediate and advanced users.

Dos and don'ts

- Build Reference topics so that they don't need to be read completely to find and retrieve specific information.
- Keep your text as concise as possible.
- It's OK to use telegram style and incomplete sentences.
- Apply a clear and consistent structure.
- When possible, layer the information.
- Add descriptive labels (subheadings) to help readers skim the text for relevant information (see *Add labels (subheadings)* 54).
- When possible, use tables and lists rather than running text.
- It's OK to provide a large number of links to related information.
- Don't provide detailed explanations. If you need to provide background information, link to a Concept topic instead (see *Writing "Concept" topics* 46).
- Don't give instructions. For this purpose, use a Task topic instead (see *Writing "Task" topics* 47).

Writing a user interface reference

When writing a Reference topic that describes the controls of a user interface:

- Begin with a brief statement describing the purpose of a dialog (when documenting software) or the purpose of a group of controls.
- If there are general principles and dependencies that affect all controls, describe these principles before focusing on the individual elements.
- Describe the controls in the following order, as applicable:
  - from the upper left corner toward the lower right corner
  - clockwise

- Use characteristic verbs that provide information about what users should do with a particular control.

✘ No: *Time period*

(Doesn't tell readers whether they need to enter something.)

✔ Yes: *Shows the remaining time period in days*
(Tells readers that this is an output field that doesn't require any action.)

or:

Asks for how many days you wish to stay
(Tells readers that they must enter some data.)

Related rules

Be parallel 32

Writing "Concept" topics 46

Writing "Task" topics 47

2.3 Writing sections

Don't mix information. Label each section with a descriptive subheading that clearly communicates what the section is about. Within the section, only cover what the subheading indicates.

Rules for writing on the section level

- *Don't mix subjects* 52
- *Add labels (subheadings)* 54
- *Writing procedures* 56
- *Writing lists* 64
- *Writing tables* 68
- *Writing warnings* 72
- *Writing notes* 76
- *Writing tips* 78
- *Writing examples* 79
- *Writing cross-references and links* 81

Related rules

Writing in general 15
Writing topics 45
Writing sentences 87
Writing words 113

2.3.1 Don't mix subjects

Long, continuous text can be extremely hard to read, especially when it covers various subjects. Long sections of continuous text also make it almost impossible to skim a text for specific information.

1 Don't mix various subjects within one paragraph. If there's a new subject, start a new paragraph.

It's perfectly OK if sometimes there's only one sentence within a paragraph. Don't combine paragraphs or add needless text only for the sake of building a "complete" paragraph.

As a general rule, start a new paragraph for each new idea, but don't start a new paragraph for each new sentence. A good paragraph length is about 2 to 5 lines.

2 Don't mix various subjects within one sentence. If there's a new subject, start a new sentence.

In procedures, don't describe more than 2 steps in the same sentence (see *Writing procedures* 56).

In warnings, only mention one cause of the hazard per sentence (see *Writing warnings* 72).

1

✘ **No:** *You can use the camera for taking pictures and for making short movies. Possible resolutions for pictures are a x b pixels and c x d pixels. Possible resolutions for movies are e x f pixels and g x h pixels. Pictures can be saved in JPG format; movies can be saved in AVI format.*

(Several subjects are mixed within this paragraph.)

✔ **Yes:** *You can use the camera to take pictures and to make short movies.*

Possible resolutions for pictures are a x b pixels and c x d pixels. Possible resolutions for movies are e x f pixels and g x h pixels.

Pictures can be saved in JPG format; movies can be saved in AVI format.

(Here, the subjects are separated into sections: what you can do with the camera (first paragraph), the possible resolutions (second paragraph), and the formats (third paragraph).)

✔ **Yes:** *You can use the camera to take pictures and to make short movies.*

Possible resolutions for pictures are a x b pixels and c x d pixels. Pictures can be saved in JPG format.

Possible resolutions for movies are e x f pixels and g x h pixels. Movies can be saved in AVI format.

(Here, the information is rearranged. The idea of the second paragraph is "pictures"; the idea of the third paragraph is "movies.")

✔ **Yes:**   *Camera A has been designed for children. Camera B has been designed for adults.*

(There's no new paragraph because the direct comparison of both models is one common subject.)

✔ **Yes:**   *Camera A has been designed for children.*

You can use the camera to:

- *take pictures*

- *make short movies*

Each function can be activated with a single key.

(Here, there are separate paragraphs because there are several independent ideas: the fact that the camera is for children, the different things you can do with the camera, and the fact that each function can be activated with a single key.)

✘ **No:**   *You can use the text import function, which can import PDF files, Word files, and XML files, to take over texts that were written with another program.*

✔ **Yes:**   *You can use the text import function to take over texts that were written with another program. You can import PDF files, Word files, and XML files.*

Related rules

Add labels (subheadings) 54

2.3.2 Add labels (subheadings)

Start each new section with a label that clearly communicates what the section is about.

In a printed document, labels often appear as subheadings or as subtitles in a margin column. In online help, labels often appear as subtitles as well, or they are clickable headings of expandable sections (toggles).

Readers benefit from the labels because the labels work as landmarks that enable them to skim a text for specific information and to decide beforehand whether to invest the time in reading a particular section.

As the author, you also benefit from the labels:

- Labels give you a constant overview of the structure of your document. You can quickly see where the best place is to add new information.

- Labels minimize the risk of mixing subjects (see *Don't mix subjects* 52). When writing, you can constantly check your text against the label. If the text doesn't fit under the label, this indicates that you should put the information somewhere else.

How to phrase a label

You can either use a short sentence or statement as label text, or you can write the label like a heading.

- Make sure that the label text is concise and easy to read.

- Make the label meaningful. The label should clearly communicate:

  - what the section covers (the subject)

  - what kind of information readers can expect (concept, task, or reference)

  - what level of detail is given (basic information for beginners or details for advanced users)

If the surrounding topic doesn't mix information types but is clearly either a Concept topic, a Task topic, or a Reference topic, labels can be very short and don't have to communicate the information type again. For example, if the surrounding topic is clearly a Task topic, it's evident that subsections also cover tasks.

If the surrounding topic mixes various information types (generally not recommended), you need to find labels that communicate the information type of each section as well. For example, you could use phrases such as "How to print ..." or "Printing ..." to indicate that a section contains a procedure.

How many sentences and paragraphs should go into one labeled section?

There's no general rule about how many labels you should add. A good average is somewhere between 2 and 7, but this largely depends on the subject.

As rules of thumb:

- Never mix different subjects within one section under the same label. If there's a new subject, start a new section.

- When in doubt, it's better to start a new section with an additional label than to have a section that's longer than half a page.

- A section may contain a single paragraph or multiple paragraphs. The paragraphs may all be of the same type (such as procedures) or of different types (such as body text, a procedure, and a warning).

Establish a consistent labeling scheme

When possible, establish a consistent way of structuring your topics and of labeling the sections within these topics. This adds consistency and parallelism (see *Be consistent* 29 and *Be parallel* 32).

For example, you could:

- always use the sections "Requirements," "Steps," and "Results" in Task topics

- always use the sections "Purpose," "Syntax," "Input," "Output," and "Parameters" in a particular kind of Reference topic

Note that it depends on your specific product and use case which labels work best.

Related rules

Don't mix subjects 52

2.3.3 Writing procedures

Procedures are the core part of user assistance. A procedure is a description of the steps that users have to follow to complete a specific task.

- A procedure typically begins with an introductory phrase that states the goal of the procedure and ends with a colon. Sometimes the introductory phrase can also be a statement, which ends with a colon or period.

- Steps are numbered. They consist of action statements, descriptions of the system's response, and related information that enable users to execute the procedure.

- The procedure typically ends with a brief description or picture of the result.

✔ **Yes:** To set the alarm time:

1. Hold down button A for at least three seconds.
   The time display now blinks and shows the currently set alarm time.

2. Press the arrow buttons repeatedly to change the alarm time.

3. Press button B to store the new alarm time.
   The new alarm time now appears at the bottom of the display after the chime symbol.

✔ **Yes:** Writing a manual involves three major tasks:

1. Planning the structure.

2. Designing the template.

3. Writing the texts.

(Strictly speaking, this is not a classical procedure but a process description. The introductory phrase is a statement, although it also states the goal (writing a manual). The steps aren't written in an imperative form.)

Basic rules

Never use body text for procedures. Always use numbered steps.

Don't use words to describe the succession of steps, such as "first," "next," "in the following step," and so on.

Don't merge multiple actions into one step.

✘ No: *To change the brightness of the screen, you first need to press the red button. The value of the current setting then appears on screen. Next, press the arrow keys to change the brightness. When you've finished, finally press the green button. The new brightness value is now set.*

✘ No: *To change the brightness of the screen:*

1. *First, press the red button. The value of the current setting now appears on screen. Next, press the arrow keys to change the brightness. When you've finished, continue with step 2.*

2. *Finally, press the green button. Congratulations! The new brightness value is now set.*

✔ Yes: *To change the brightness of the screen:*

1. *Press the red button. The value of the current setting appears on screen.*

2. *Press the arrow keys to change the brightness.*

3. *Press the green button. The new brightness value is now set.*

Talk to readers clearly and directly and use the imperative.

✘ No: *The next step consists of pressing the Stop button.*

✘ No: *One must press the Stop button now.*

✘ No: *Now the Stop button should be pressed.*

✘ No: *Now it's necessary to press the Stop button.*

✘ No: *Now you need to press the Stop button.*

✘ No: *Now you must press the Stop button.*

✘ No: *Now, please press the Stop button.*

✔ Yes: *Press the Stop button.*

Describe only the simplest, most common way of doing things. When documenting software, describe keyboard shortcuts in a separate reference topic of shortcuts.

✘ No: *Click File > Save. Alternatively, you can also press [Ctrl]+[S].*

✔ Yes: *Click File > Save.*

If a procedure involves any hazard, add a warning *directly before the step that involves the hazard* (see *Writing warnings* 72). Don't add the warning after the step.

If a warning relates to the procedure as a whole, add the warning at the beginning of the procedure.

Keep procedures as short as possible. If a procedure has more than 7 to 10 steps, consider splitting the procedure. If it's not possible to split the procedure, add labels (subheadings) that group steps into subtasks or that mark certain milestones in the procedure.

Mention all prerequisites

Don't assume that the correct conditions are already in place.

- With products that involve navigation through menus, don't assume that users already are in the right place to start a procedure. They could be *anywhere* within the on-screen menu when reading your instructions.
- In a service manual, don't assume that users have already disassembled certain components or have already completed other preparatory steps.
- Bear in mind that users may have done something that will prevent the procedure from working.

Consequently, clearly state all prerequisites at the beginning of the procedure. In addition, if any special tools are required, list these tools so that users can fetch them in advance.

Often, the best way of stating the prerequisites is to make the preparation the first step of the procedure ("1. Make sure that you have ... available.", "1. Prepare ....").

Alternatively, you can formulate the prerequisites as a condition that you put before the steps ("You must have ... before ...."). If you have a large number of prerequisites, put them into a table or list.

If you have various similar procedures, state the prerequisites in a consistent, parallel way. For example, always use a table, and always use the same structure within this table.

Mention the purpose of a step if this is information is helpful

Don't hesitate to explain the purpose of a step. People are often more willing to execute a particular step and perform it better when they understand why they need to do it. Half a sentence is often enough.

✖ No:
...

5. Fix the bolt with a small piece of adhesive tape.

...

✔ Yes:
...

5. To prevent the bolt from falling down, temporarily fix it with a small piece of adhesive tape.

...

Describe the system's responses

To give users the possibility to control the success of their actions, describe not only the users' actions but also what happens as a result of critical actions. Feedback is especially important if many of your readers are inexperienced users.

Sometimes you can embed the information about the response right into a step. If that's not possible, add a line break, and then add a second sentence.

✔ Yes: 1.*Click Options to display more fields.*

         2.*Select Landscape.*

✔ Yes: 1.*Click Options.*
           The dialog box expands.

         2.*Select Landscape.*

If the change of state has important negative implications, include these in a note or important note (see *Writing notes* [76]).

At the end of the procedure, describe or show the final result.

First tell where to act, then tell what to do

Make it easy for readers to follow your instructions step by step. Get your words into the same order that users must follow both mentally and physically.

Match the order of words with the sequence of steps that users must take to identify an object. Always tell users where the action takes place before describing the action to take.

✘ No: *Click Save in the Options window.*

✔ Yes: *In the Options window, click Save.*

         (Users must first find the **Options** window, then find and click the **Save** button.)

Combining steps

Usually, create one sentence for each step. This breaks down a complex task that's difficult to perform into a number of simple subtasks that are easy to perform. In addition, it helps you to bring the information into the right order.

Avoid "do-this-after-you-have-done-that" statements.

✘ No: *To write a user manual, enter the text into the previously opened word processor, according to the structure of the outline that you've set up before beginning to write.*

✔ **Yes:** *To write a user manual:*

1. *Set up an outline.*

2. *Open your word processor.*

3. *Enter the text according to the structure of your outline.*

It's OK to combine two steps if:

- both steps are short
- both steps occur in the same place or affect the same component

✔ **Yes:** *On the **Tools** menu, click **Options**, and then click the **Edit** tab.*

✔ **Yes:** *Insert the probe holder and turn it clockwise until it snaps into place.*

Note that the word *then* is not a coordinate conjunction and thus cannot correctly join two independent clauses. Always add "and."

✘ **No:** *On the **Tools** menu, click **Options**, then click the **Edit** tab.*

✔ **Yes:** *On the **Tools** menu, click **Options**, and then click the **Edit** tab.*

Handling branches and alternatives

Make sure that each numbered step contains at least one action for every reader. For this reason, optional steps shouldn't have their own number.

Always start with the most common (default) action, followed by the alternative action.

For choices within one procedure step, use a list.

✘ **No:** ...

7. *If you live in Europe, enter EUR.*

8. *If you live in America, enter AME.*

9. *If you live in any other part of the world, enter GLOB.*

...

✔ **Yes:** ...

7. *Enter your region code:*

   - *If you live in Europe, enter EUR.*
   - *If you live in America, enter AME.*
   - *If you live in any other part of the world, enter GLOB.*

...

(Note: This example assumes that the product is mainly sold in Europe, so Europe is the first item in the list.)

Put the if-condition at the beginning of the sentence so that readers for whom the condition isn't fulfilled can skip the rest of the sentence.

✘ **No:** *Open the cover and check that a color cartridge has been inserted if you want to print in color.*

✔ **Yes:** *If you want to print in color, open the cover and check that a color cartridge has been inserted.*

Handling loops

When users have to repeat one or more steps multiple times, you often can't avoid having a loop within your procedure.

Always loop to something recognizable, like a step number.

✔ **Yes:** ...

> *7. Repeat steps 3 through 6 until ....*

...

Note:
Looping to a step number is usually the most user-friendly solution because this clearly identifies the beginning of the loop at a glance. However, bear in mind that this solution is also very error-prone. If you add or remove a step from the procedure later, the step numbers will change. You must then also change the loop information, which you can easily forget to do.

Capitalization and punctuation in procedures

Precede all procedures with colons, regardless of whether the text before the colon is a complete sentence or partial sentence.

Note:
Microsoft recommends that you don't use a colon or any other punctuation here. If you want to create documents that mimic Microsoft user assistance, omit the colon.

Use sentence-style capitalization for each step.

Add a period at the end of each step. Don't use exclamation points.

Alternatives to step-by-step procedures

Sometimes, step-by-step procedures aren't the right means to communicate a task.

Flowcharts:

If a procedure is essentially a decision-making process rather than a straightforward action, consider using a flowchart. A flowchart graphically shows the decision points and the pathways that users must follow through the branching logic of the procedure.

In particular, flow charts are often a good choice for troubleshooting and repair procedures. If the results of the decision tree are recommended actions, you can link to detailed step-by-step procedures from the flow chart.

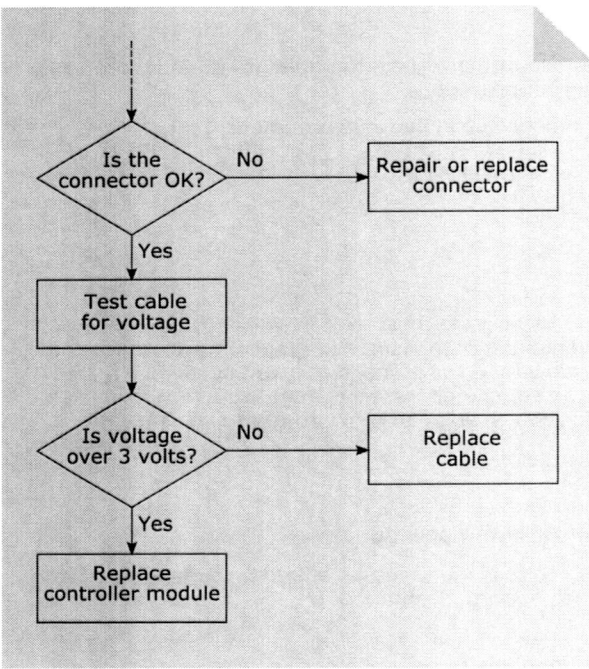

Play script format:

If a task involves various stakeholders, a play script format can be a good choice. Often, such tasks are administrative tasks within an organization.

Each person's actions are listed separately, so each person clearly sees which steps must be done by him or her, and also sees the global context of these actions in the whole procedure.

One column lists the actors; the other column lists the corresponding steps.

Person in charge	Step
Administrator	**1.** The first step.
	2. The second step.
Operator	**3.** The third step.
Technician	**4.** The fourth step.
	5. The fifth step.
	6. The sixth step.
Administrator	**7.** The seventh step.
	8. The eighth step.

Related rules

Be parallel 32

Writing "Task" topics 47

Writing lists 64

2.3.4 Writing lists

Lists present small portions of information in a well-structured way. This helps readers to scan a text quickly and to find information easily.

Lists also help you, the author, because they organize your thoughts.

There's no strict sequence in the items of the list.

List items are bulleted, not numbered. Don't confuse lists with procedures (see *Writing procedures* 56).

Basic rules:

1 If it's not obvious from the heading what a list is about, begin the list with an introductory phrase. The introductory phase can either be a complete sentence or a sentence fragment.

Don't build a list and its introductory phrase as one continuous sentence.

- Don't continue the introductory sentence through the list.
- Don't use words and phrases such as "and," "as well as," or "or" to connect the items of a list.

There are several reasons why you shouldn't build lists that are actually one big sentence:

- You should support readers who don't read the complete text, so it's important that each list item is comprehensible on its own.
- Lists that are actually one big sentence can be very hard to translate into foreign languages.
- Adding items to the list, deleting items from the list, or reversing the order of items requires modifications also in other list items, which is often forgotten.

2 Keep all list items as short as possible. List items may be full sentences or sentence fragments; however, avoid mixing full sentences with sentence fragments within the same list.

Build all list items in a parallel, consistent way (see *Be parallel* 32).

1

✖ No: *You may use*

  a CD.

  a DVD, or

  a USB drive

to save your data.

✔ **Yes:**   You can save your data on:

  ▪ CD

  ▪ DVD

  ▪ USB drives

2

✘ **No:**   In a document, white space is important for the following reasons:

  ▪ visual separation of sections

  ▪ attention focus

  ▪ white space breaks content into smaller chunks

✔ **Yes:**   In a document, white space is important because it:

  ▪ visually separates sections

  ▪ focuses attention

  ▪ breaks content into smaller chunks

Punctuation and capitalization in lists

Usually, regardless of whether an introductory phrase is a complete sentence or a sentence fragment, place a colon after the introductory phrase. The colon clearly signals the beginning of a list and creates anticipation.

Note:
If the introductory phrase is a full-sentence statement, a colon is sometimes too emphatic. It's OK to use a period then.

✔ **Yes:**   With DemoSoft, you can:

  ▪ write texts

  ▪ plan projects

✔ **Yes:**   You can use DemoSoft to write several types of documents:

  ▪ letters

  ▪ reports

  ▪ books

If *all* list items are complete sentences, use sentence-style capitalization and punctuation. A list item is considered a complete sentence if, removed from the context of the list, it could stand on its own as a sentence.

✔ **Yes:** *To avoid loss of data:*

    ▪ *Make a daily backup copy of all files.*

    ▪ *Use the latest security software and update this software regularly.*

If *all* list items are sentence fragments:

- Use lowercase for the initial word of each list item (except for names).

  Many style guides also suggest capitalizing the first word in this case. This is also OK. However, no matter which format you prefer, use it consistently throughout your whole document.

- Don't use periods, semicolons, or commas to end list items if these list items aren't full sentences. Adding punctuation isn't strictly wrong, but it adds unnecessary clutter.

- Don't use a period after the last list item.

- Don't use a period even if a list item completes the introductory clause of the list (not recommended).

✘ **No:** *With DemoSoft, you can:*

    ▪ *write texts,*

    ▪ *plan projects.*

✔ **Yes:** *With DemoSoft, you can:*

    ▪ *write texts*

    ▪ *plan projects*

If *some* list items are complete sentences and other list items in the same list are sentence fragments (not recommended):

- Start each list item with a capital letter.

- End each list item with a period.

If essentially all list items are sentence fragments, but one or more of these fragments are followed by a complete sentence, use sentence-style capitalization and punctuation for all list items. As an alternative, you can also use semicolons.

✔ **Yes:** *With DemoSoft, you can:*

    ▪ *Write texts. (You must have the extended text module installed to use advanced editing features.)*

    ▪ *Plan projects.*

alternatively:

With DemoSoft, you can:

- *write texts; (you must have the extended text module installed to use advanced editing features)*

- *plan projects*

Related rules

Be parallel 32
Writing procedures 56

2.3.5 Writing tables

Tables are a way to:

- structure information
- visualize relationships
- present a large amount of uniform information in a condensed way

Whenever items can be grouped by at least two characteristics, you can present them as a table.

Within tables, you can use all other standard elements, such as lists, warnings, notes, and figures.

Tables are usually more appropriate for Reference topics than for Concept topics and Task topics.

Advantages of tables:

- Tables can present a large amount of information in a small space.
- Tables can replace long sections of hard-to-read text.
- Readers can find specific information quickly.
- Readers can easily compare data.
- Tables are easy to write and easy to translate.

Disadvantages of tables:

- For some audiences, tables are hard to interpret.
- Tables can sometimes take up more space than regular text.
- Space in table cells is limited, especially if the table has many columns.

Basic rules and tips

- Add as few columns as possible. If the layout of a table is awkward, consider switching rows and columns. If a table has too many rows or columns, consider breaking the table into two or more tables.
- Keep all texts within the table short.
- If the type of information that each column contains is obvious, omit the heading row.
- Within a column, keep entries parallel (see *Be parallel* 32).
- For numbers, use the same unit of measure in all table cells.

- For numbers, use the same number of decimals in all table cells.

Arranging information within a table

Arrange rows and columns so that finding information seems "natural."

- When designing table rows and column heads, anticipate the readers' questions.
- Put the information that readers are most likely to search for into the first column.

✖ No:

Keyboard shortcut	Purpose
Ctrl+A	select all
Ctrl+C	copy selection to clipboard
Ctrl+F	find & replace
Ctrl+O	open file
Ctrl+P	print active document
Ctrl+S	save active document
Ctrl+V	paste from clipboard
Ctrl+X	cut selection and copy it to clipboard
...	...

(Users aren't interested in the fact what Ctrl+S does, for example. Users would rather want to know which keyboard shortcut they can use to save a file.)

✔ Yes:

Purpose	Keyboard shortcut
Handling files	
open file	Ctrl+O
save active document	Ctrl+S
...	...
Editing documents	
select all	Ctrl+A
find & replace	Ctrl+F
...	...
Working with the clipboard	

copy selection to clipboard	Ctrl+C
cut selection and copy it to clipboard	Ctrl+X
paste from clipboard	Ctrl+V
...	...
Printing	
print active document	Ctrl+P
...	...

Within the first column, sort items numerically or alphabetically, or find another structure that reflects the mental model, tasks, goals, and priorities of the readers.

When possible, place those items closely together that are related or that readers are most likely to compare.

Tip:
Some authoring tools allow you to provide sortable tables for online help. When a reader clicks a column heading in the sortable table, the table is sorted according to the contents of this column.

Referring to tables

You don't have to introduce a table with a special sentence.

When possible, design and arrange your content so that you don't have to explicitly mention the existence of a table.

Design the column headings so that it's immediately apparent what the table shows.

✘ **No:** The thermostat lets you choose the desired temperature. The following table shows you which settings are available and which settings you should use.

Setting	Use
1	unheated rooms, such as stairways or storage rooms
2	moderately heated rooms, such as bedrooms or kitchens
3	warm rooms, such as living rooms or bathrooms

✔ **Yes:** The thermostat lets you choose the desired temperature for each room.

For ...	Choose setting ...
unheated rooms (stairway, storage room)	1
moderately heated rooms (bedroom, kitchen)	2
warm rooms (living room, bathroom)	3

Only point out a table if you're afraid that a significant proportion of readers might otherwise ignore the table.

Place the reference to the table in the paragraph that immediately precedes the table.

Use the word *following* to refer to the position of the table. Avoid the word *below* because if there is a page break after the table in printed content, the table isn't actually "below" anymore.

✘ No: The table below lists ....

✔ Yes: The following table lists ....

Capitalization and punctuation in tables

In row headings and column headings, use sentence-style capitalization (see *Capitalization of headings* 165).

Handle the capitalization and punctuation of text in regular table cells like the capitalization and punctuation in lists (see *Writing lists* 64).

Related rules

Be parallel 32

2.3.6 Writing warnings

Warnings alert users to potential hazards to people or products.

Particular warnings are also often required for legal reasons.

> ⊕ **Important:** Comply with the appropriate safety standards, depending on your product and on the countries where your product is sold. These rules have priority over the general recommendations given here. For example, if warnings follow the ANSI Z535.4 and ISO 3864-2 standards, you must add a specific safety symbol, depending on the particular hazard.

It's important not just to tell users what to do or not to do, but to help them understand *why* they should take some particular precautionary measures. Only if users are aware of the reasons for a measure, and of the personal consequences and implications of not following the advice, will they take the warning seriously. For this reason, each warning must provide the following information (in this order):

- Safety alert symbol.
- Signal word that indicates the severity of the hazard: *Caution*, *Warning*, or *Danger* (see the following section on the types of warnings).
- Information on what kind of danger exists, and on where the danger comes from.
- Information in what can happen to the user, to other people, to the product, and to other things.
- Information on how the danger can be avoided altogether, or on how the risk can at least be minimized.

✔ **Yes:** ⚠ *WARNING*
Moving parts can snag and pull.
May cause severe injury.
Do not wear loose clothing or jewelry, and pull back long hair in a hairnet or ponytail while operating the machine.
Do not operate the machine with the guard removed.

ANSI Z535.4 and ISO 3864-2 standard warnings look like this:

Safety symbol

Safety alert symbol and signal word

⚠ WARNING

Moving parts can snag and pull.
May cause severe injury.

▶ Do not wear loose clothing or jewelry and pull back long hair in a hairnet or ponytail while operating the machine.

▶ Do not remove the guard.

Kind of hazard

Consequences of not avoiding the hazard

Measures to avoid the hazard and to minimize the risk

Basic rules

- Always place the warning *directly before* the step that's dangerous or causes a danger. Users mostly follow instructions step by step. If the warning comes after a step, it may come too late. If the warning comes at the beginning of the topic, long before the instruction, readers might not read it at all or may have forgotten it when they come to the dangerous step.

- If a warning is independent of particular procedures and refers to working with the device in general, put it into a special safety instructions topic at the beginning of the document.

- Keep warnings short and to the point. The optimal length is one sentence. Avoid warnings that are longer than 3 sentences. Consider splitting longer warnings into 2 separate warnings. If explanations or instructions are needed, put them into other sections after the caution.

- When necessary, don't hesitate to repeat something that you have already said somewhere else.

- Invest much effort in writing clearly and simply (see *Writing sentences* 87 and *Writing words* 113).

- Invest much effort in writing so that readers who speak the document's language only as a second language can precisely understand the hazards.

- Address readers directly using the imperative form (see *Talk to the reader* 21). Don't use the passive voice (see *Use the active voice* 39).

- Use appropriate vocabulary to point out the possible consequences of disregarding the warning.
- Don't use contractions; in particular, don't use contractions with *not*. For example, use *do not* instead of *don't*, use *must not* instead of *mustn't*, and so on.
- Write a warning only if an action involves any real danger for people or things. To provide information that helps to prevent mistakes that don't result in injury, death, or damage, write a note of the type "important note" instead (see *Writing notes* 76).
- Don't use warnings in place of tips. Write a tip instead (see *Writing tips* 78).

Types of warnings: CAUTION, WARNING, DANGER

The signal word at the beginning of the warning indicates the severity of the danger. To make it particularly emphatic, it's often written in capital letters.

- A warning that begins with the signal word **CAUTION** indicates a hazard that, if not avoided, *might* result in *minor* or *moderate* injury. A warning with the signal word **CAUTION** can also refer to a situation that could damage or destroy the product or the users' work. If a hazard doesn't involve any risk for people but only for things, the warning symbol is often omitted or the signal words **SAFETY INSTRUCTIONS** are used instead of **CAUTION** (ANSI Z535.6).
- A warning that begins with the signal word **WARNING** indicates a hazard that, if not avoided, *could* result in *death* or *serious injury*.
- A warning that begins with the signal word **DANGER** indicates a hazard that, if not avoided, *will* result in *death* or *serious injury*.

✔ **Yes:** ⚠ *CAUTION*
The lamp may get very hot.
To avoid skin burns, wait a few minutes before exchanging a defective lamp, or wear protective gloves.

⚠ *CAUTION*
Magnetic field.
May damage electronic components.
Keep electronic components away.

⚠ *CAUTION*
Formatting the disk deletes all data that have previously been stored on this disk.
Make a backup copy and store it in a safe place before you proceed.

✔ **Yes:** ⚠ *WARNING*
Ultraviolet radiation.
May cause irreversible eye damage.
Do not look directly at the UV lamp.

⚠ *WARNING*
Magnetic field.
Can be harmful to pacemaker wearers.
Maintain a distance of at least 30 cm (12 in.) from the equipment.

✔ **Yes:** ⚠ *DANGER*
High voltage.
Contact will cause electric shock, burns, or death.
Disconnect all power sources before removing the panel.

When to use exclamation points

Usually, you don't need to use exclamation points within warnings because the warning symbol and the signal word already add enough emphasis.

However, if the warning symbol and the signal word are missing (not recommended), do add an exclamation point. You can also use an exclamation point if you want to put some special emphasis on a particular statement.

✔ **Yes:** ⚠ *WARNING*
Hot surface.
Risk of burns.
Do not touch!

Related rules

Exclamation points 182
Writing notes 76
Writing tips 78
Writing examples 79

2.3.7 Writing notes

There are two types of notes:

- A **standard note** supplies neutral or positive information that *supplements* the main text. A standard note can also supply information that only applies to special cases.

- An **important note** calls the readers' attention to an *important* point about typical obstacles. These obstacles may prevent users from obtaining the desired results, but they don't cause injury, damage, or loss of data. If there's any risk of injury, damage, or loss of data, add a warning instead (see *Writing warnings* 72).

  Important notes can also provide information that's required for legal reasons.

Basic rules:

- Use notes sparingly. Notes create an abrupt interruption in text. What's more, notes lose their effectiveness if they appear too often. If your text is cluttered with notes, you should urgently reorganize the information.

- Don't use notes as dumping grounds for information that should go somewhere else.

- Don't use a note for a warning. Write a warning instead (see *Writing warnings* 72).

- Don't use a note for a tip. Write a tip instead (see *Writing tips* 78).

- At the beginning of a standard note, add the word *Note* or an icon that shows that the information is a note. At the beginning of an important note, add the word *Important* or an icon that shows that the information is an important note.

- Keep notes short and to the point. The optimal length is one sentence. If a note is longer than 3 sentences, consider transforming it into a subsection with an appropriate subtitle.

✔ **Yes:** *Note:*
Don't use notes as a place for information that doesn't fit anywhere else.

✔ **Yes:**
❶ *Important: Don't forget to check the battery status of the device before taking it on a trip. A full battery will last 8 to 10 hours.*

✔ Yes:

> **❶ Important:** *Using the lawn mower may be annoying for your neighbors. Follow the regulations of your local community as to when using lawn mowers is permissible.*

Related rules

Writing warnings 72

Writing tips 78

Writing examples 79

2.3.8 Writing tips

Tips help users:

- to perform a task more efficiently
- to perform a task more effectively
- to perform a task more conveniently
- to find additional applications for a feature or product

Tips aren't required to perform a particular task.

Basic rules:

- Don't misuse tips for advertising. Only add tips that provide some real benefit for the users, not for you as the manufacturer.
- Don't include any information that's necessarily required. It must be optional to read a tip.
- Don't use a tip for a warning. Write a warning instead (see *Writing warnings* 72).
- You can use a tip to provide information on how to produce better results, but don't use a tip for information that helps to prevent mistakes. Write a note of the type "important note" instead (see *Writing notes* 76).
- At the beginning of the tip, add the word *Tip* or an icon that shows that the information is a tip.
- Keep tips as short as possible. The optimal length is one sentence. If a tip is longer than 3 sentences, consider transforming it into a subsection with an appropriate subtitle.

✔ **Yes:** *Tip:*
        To make tips emphatic, avoid the passive voice. Talk to the reader directly.

Related rules

Writing warnings 72
Writing notes 76
Writing examples 79

2.3.9   Writing examples

> Examples:
>
> - set a context for a feature or task
> - make your documents easily understandable
> - make your documents more memorable
>
> Examples can be small or large. An example can be just a short comment in parentheses, or it can be a complete topic with several sections, including lists, tables, pictures, notes, and warnings.
>
> Use examples frequently, but only use them if they actually add some value to what you've said in the surrounding text.

Basic rules

- Make it clear that you're giving an example, not a rule.
- Find practical examples that come close to the actual work of the readers.
- When possible, find a common scenario for your examples. Once readers know the example scenario, they can then better concentrate on your message rather than on the details of the example.
- Keep the example as simple and stupid as possible. Leave out all needless details. Simplify or blur everything that doesn't relate to the key message.
- Make sure that it's clear where the example ends and the general information continues.

Making examples multicultural

When writing examples, it's particularly important that you take into account cultural differences among the countries in which your product is sold and used. If you don't, users in other regions of the world may not be able to understand your examples, or they might even be offended by what you show or say.

In particular:

- If you need to mention or show names of people, use names of different ethnic groups, and mix genders.
- If you need to mention countries, cities, or places, choose places from different regions.
- Vary locally specific things like currencies, addresses, phone numbers, e-mail addresses, URLs, and so on.

- Avoid mentioning politics, religion, sports, traditions, holidays, and all sorts of special events.
- Avoid relating to particular social, legal, and business standards and practices.

Developing an example theme

If it's possible, choose a theme for examples that you can use throughout your document:

- You can often reuse certain parts of the example.
- Readers who are already familiar with the example theme can fully focus on your message rather than on the details of the example.

When possible, also use the data from the example in your pictures—even if the pictures are not part of the examples. Again, this helps both you, the author, and your readers:

- You can use the same demo setting to make the pictures.
- Your readers see familiar data that relate to the given examples.

Example: In the documentation for some collaborative text processing software, the theme could be that a small team wants to write a guideline on how to handle complaints from customers. As part of your theme, you could establish a name for the company as well as for the people involved, and you could then carry those names through all the examples in your document.

2.3.10   Writing cross-references and links

When writing cross-references and links, be aware of their advantages and disadvantages.

Advantages of cross-references and links are:

- they make it easy to find related information
- they make it possible to keep topics short; you can give optional or related information in another, linked topic

Disadvantages of cross-references and links are:

- they attract attention and interrupt the flow of reading
- they require a decision from readers whether or not to follow the link
- if readers follow the link, they often do so without having read the complete topic; if they don't return, they miss what comes after the link
- if readers don't follow the link, they might also miss some important information—or they might feel as though they are missing something

Readers are constantly faced with a dilemma: Should they yield to the alluring call of a link? Should they leave the current topic, or should they read on?

If readers decide to click a link but the topic that opens up doesn't contain the expected information, they feel that the document doesn't answer their questions and eventually stop reading.

Basic rules

The most important basic rules are:

- Use cross-references and links sparingly.
- Always link to a particular topic or page. Don't add references such as "... is described elsewhere" or "... can be found in another help topic."
- Choose a link text that's clear in its meaning even when a user doesn't read the surrounding text. Formulate link texts so that the content and information type of the target topic are clearly recognizable.
- Keep link texts simple and concise.
- When readers click a link, something new and unknown awaits them. Give them a sense of security by making sure that at least one keyword from the link text corresponds to the title of the subject invoked. This will affirm to them that they've landed in the right place, which creates confidence and

contributes to a positive user experience.

- Make links unobtrusive. When possible, integrate links seamlessly into the sentence so that the sentence doesn't change because of the link.
- Place links preferably at the end of sentences, paragraphs, and topics. In these places, they don't interrupt the readers.
- If you create printed manuals and online help from the same text base (single source publishing), avoid using media specific terms like *topic* or *chapter*.
- Embed only very important links into the running text. Move the majority of links out of the text into a separate navigation zone. This approach has a number of advantages:
  - readers' attention isn't diverted from the text that they're reading
  - formulation is significantly simpler, especially if you want to generate a printed manual and online help from the same text base (single source publishing)
  - administration is significantly simpler (this depends on the authoring tool)
  - translation is significantly simpler

As a rule of thumb, include only those links in the main body of the text that really justify any interruption in the flow of the text. Ask yourself the following question: If you weren't writing but talking in a face-to-face conversation, would you interrupt yourself to draw the other person's attention to some additional information? If the answer is "yes," the link is probably justified in the text. If the answer is "no," move the link to a separate list of related topic links below the topic's body text.

Cross-references in printed manuals vs. links in online help

In a printed manual, a cross-reference needs a page number. Most authoring tools add the page number automatically when you generate the final document. To keep cross-references as short as possible, don't include any chapter numbers.

✖ **No:** ... chapter 2.3.4 (page 75)

✖ **No:** ... in chapter 2.3.4 on page 75

✔ **Yes:** ... (page 75)

✔ **Yes:** ... on page 75

If you want to generate a printed manual and online help from the same text base (single source publishing), also don't include any words that are typical for a printed manual or for online help in your source files, such as the word "page." Powerful single source authoring tools are capable of adding these terms automatically as required.

After production, the final results in your documents should look something like this:

✔ **Yes:** Online help: *You can find additional information about the graphics format under The right format for your project.*

✔ **Yes:** Printed manual: *You can find additional information about the graphics format under "The right format for your project" on page 225.*

✔ **Yes:** Online help: *Before you can use this function, you need an additional license key.*

✔ **Yes:** Printed manual: *Before you can use this function, you need an additional license key (see "Acquiring additional license keys" on page 13).*

Making link texts clear

Give readers some decision aid:

- Where will a link take them?
- What kind of information can they expect there?
- For which applications do they need this information?
- Will there be plenty of information, or just the bare minimum?
- Will the linked information be easy or difficult to understand?
- Will there be practical examples or abstract theory?

The more of these questions you can answer even before a reader clicks the link, the better.

Because many readers just scan online help text briefly, it's generally not enough if the information about the link is given in the accompanying text. Make it present in the link text itself.

✘ **No:** *Click here for more information.*

✘ **No:** *Click here for more information.*

✘ **No:** *Click here for the results of our usability study.*

✔ **Yes:** *Most users prefer links at the end of a topic. This is confirmed by the results of our usability study on navigation (3 pages).*

(Even if a reader doesn't read the complete sentence, this link text clearly communicates what the link target is about. This outweighs the disadvantage that this link text is longer than the link text in the previous examples.)

Use short phrases such as the popular "(more ...)" only if they're located near a clear title, for example, in a table or tabular listing.

✘ No: *Choose whether you want to write a letter, create a presentation, or plan a project (more ...).*

✔ Yes: *You can use the program*

- *to write letters (more ...)*
- *to create presentations (more ...)*
- *to plan projects (more ...)*

Mention any access restrictions immediately after the link but don't make this part of the link text.

✔ Yes: *You can also send reports by email (Professional Edition only).*

Making links unobtrusive

Every link draws the attention of the reader away from the actual content of the topic:

- The layout stands out prominently (underlining).
- Readers are anxious about missing important information by not following the link.

In addition, sentences often are longer and syntactically more complex if there's a link within them.

The ideal link is integrated into the text in such a way that the sentence would read just as if the link wasn't there. Try to write the link so that it only contains its content, but no extra navigational information.

✘ No: *To learn how to enter text, follow the link to the topic Using the On-Screen Keyboard.*

✔ Yes: *To enter text, use the on-screen keyboard.*

Don't advertise links. There's no need to explicitly refer to a link in the text—it stands out well enough on its own.

✘ No: *Click here to start the tutorial.*

✔ Yes: *You can find a step-by-step introduction in the Tutorial.*

Linking to pictures and tables

When possible, avoid linking to pictures and tables altogether. Try to place pictures and tables so that it's clear which picture or table the text relates to. In particular, avoid linking to pictures and tables that are in other topics.

If you can't avoid linking to a picture or table, usually the best solution is to add the plain link at the end of a sentence.

✘ **No:**   *Insert the batteries as shown in figure 4 on page 23.*
✔ **Yes:**   *Insert the batteries (fig. 4 on page 23).*

2.4 Writing sentences

Form sentences that are easy to process for the human brain.

Rules for writing on the sentence level

- *Make short sentences* 88
- *Put the main thing into the main clause* 91
- *Avoid parentheses and nested sentences* 92
- *Feel free to start sentences simply* 94
- *Feel free to end sentences simply* 95
- *Watch the order of words* 96
- *Watch the position of modifiers* 99
- *Feel free to repeat a word* 101
- *Use redundancies purposefully* 103
- *Add syntactic cues* 104
- *Be clear about what you're referring to* 106
- *Use "then" with care* 108
- *Use "and" with care* 109
- *Don't use "and/or"* 110
- *Don't use "(s)"* 112

Related rules

2.4.1 Make short sentences

Follow the rule "One idea, one sentence."

Don't put too much information into one sentence.

Long sentences make it difficult to understand a text because they consume a lot of short-term memory.

It's not so much of a problem if you have just one long sentence. It's a big problem, however, if you have a succession of many long sentences.

Instructions (procedures) need especially short sentences because users must both read and act almost simultaneously, which limits their short-term memory. Never describe more than two actions within one sentence.

Write so that your reader has to read each sentence only once. Most readers aren't willing to read a sentence a second time.

- If readers who didn't understand a sentence *do* read a sentence a second time: They'll be angry that you've stolen their precious time.

- If readers who didn't understand a sentence *don't* read a sentence a second time: They'll miss an important piece of information. Maybe they even fail to use your product correctly.

Short sentences force you, the author, to be clear. This can be hard when writing a text, but it's essential for your readers. If you can't shorten all sentences, even occasional shorter sentences are helpful.

What's a good sentence length?

As a rule of thumb, 10 to 15 words per sentence is a good *average* sentence length.

In normal text, individual sentences with subordinate clauses may be up to about 25 words long.

In instructions, the longest sentences shouldn't be longer than approximately 20 words.

Tip:
When in doubt, read your own sentence. If you can't repeat the sentence from memory, it's too long.

There may be additional factors that further limit the acceptable maximum sentence length:

- Users with a low educational level need shorter sentences than users with a high educational level.

- Users who speak the document's language as a second language need shorter sentences than native speakers.
- Users who have little time to read a text need shorter sentences—especially in stressful and dangerous situations. (Example: Instructions on a fire extinguisher.)
- Users who are standing in a hot, noisy production hall need shorter sentences than users who are sitting in a quiet office or at home.

Tips for splitting long sentences

- Search for the words *and* and *or*. Often, these words signal the easiest and most sensible places to split a long sentence.
- Convert comma-separated lists into bulleted lists (see *Writing lists* 64) or tables (see *Writing tables* 68).
- Convert "if-then" structures into bulleted lists.
- Convert sentences that describe multiple actions into a step-by-step procedure (see *Writing procedures* 56).

✘ No: *You can use the program to write letters and you can use it to display and print presentation slides.*

✔ Yes: *You can use the program to write letters. You can also use the program to display and print presentation slides.*

✔ Yes: *You can use the program:*

- *to write letters*
- *to display presentation slides*
- *to print presentation slides*

✘ No: *The manual is outdated, it was written 10 years ago by a trainee.*

✔ Yes: *The manual is outdated. It was written 10 years ago by a trainee.*

✘ No: *You have the option to print all pages, to print a range of pages, or to print the current page only.*

✔ Yes: *You can print either:*

- *all pages*
- *a range of pages*
- *the current page only*

✘ No: *Press button A if you want to perform action A, press button B if you want to perform action B, press button C if you want to perform action C, or press button D if you want to perform action D.*

✔ Yes: *Press one of the following buttons:*

To ...	press ...
(action A)	*button A*
(action B)	*button B*
(action C)	*button C*
(action D)	*button D*

✘ No: *To print the report, first open the report file, then choose **File** > **Print**, and finally click **OK**.*

✔ Yes: *To print the report:*

1. Open the report file.

*2. Choose **File** > **Print**.*

*3. Click **OK**.*

Related rules

Be concise 26

Put the main thing into the main clause 91

Avoid parentheses and nested sentences 92

2.4.2   Put the main thing into the main clause

Many writers alternate simple sentences with more complicated sentences that consist of a main clause plus one or more subordinate clauses. The advantage of this style is that then the text doesn't sound monotonous and boring. The disadvantage, however, is that you trade in variety for clarity.

Using sentences with subordinate clauses only makes sense in longer texts.

Don't use sentences with subordinate clauses at all:

- for the key information
- for things that are particularly difficult to understand

Use subordinate clauses only for those facts that are easy to understand and of minor importance.

Put the main fact into the main clause, not into a subordinate clause.

Position the key information at the beginning of the sentence.

✘ No: The PIN, which must not be noted on the credit card, consists of 4 digits.

✔ Yes: The PIN consists of 4 digits. Don't note the PIN on the credit card.

✘ No: Finally, you need to press the red button, which triggers the explosion.

✔ Yes: To trigger the explosion, press the red button.

Related rules

Make short sentences [88]

Avoid parentheses and nested sentences [92]

2.4.3 Avoid parentheses and nested sentences

Parenthetic remarks and subordinate clauses nested into a main clause make it difficult to understand a sentence. Readers must:

- memorize the beginning of the sentence
- read and mentally process the parentheses
- recall the memorized beginning of the sentence and combine it with the rest of the sentence

1 Avoid parentheses and nested sentences, regardless of whether you put them inside commas, dashes, or parentheses. If the parenthetical remark is actually important, create a separate sentence. If the parenthetical remark isn't important, omit it.

2 Parentheses are OK for:

- introducing acronyms
- adding units of measure
- referring to numbered callouts in images

3 If you have a good reason to use a nested sentence, watch the punctuation. Sometimes there are commas within one sentence that have different roles. For example, you may have a parenthetical remark that includes a comma-separated list. When two separate functions are nested inside of each other, replace one set of commas with semicolons, parentheses, or dashes.

1

✘ **No:** *Many popular programs, for example, office suites, image editors, web browsers, and email clients, still come with poor documentation.*

✔ **Yes:** *Many popular programs still come with poor documentation. These include office suites, image editors, web browsers, and email clients.*

2

✔ **Yes:** *Prefer direct current (DC) if you want to ….*

✔ **Yes:** *Enter the size (in mm).*

✔ **Yes:** *Use the emergency switch (1) only if ….*

✘ No: *The names of user interface controls, such as menu items, tab cards, buttons, list boxes, etc., are printed in italics.*

(Here, the outer commas are used for the parentheses; the inner commas are used for the list of interface controls. This isn't wrong but makes the sentence difficult to read.)

✔ Yes: *The names of user interface controls (such as menu items, tab cards, buttons, and list boxes) are printed in italics.*

✔ Top: *The names of user interface controls are printed in italics. Typical user interface controls are menu items, tab cards, buttons, and list boxes.*

Related rules

Make short sentences 88

Put the main thing into the main clause 91

that / which 252

2.4.4 Feel free to start sentences simply

There's no rule that prohibits you from starting a sentence with short words like *but*, *so*, *also*, *because*, *therefore*, *thus*, and so on.

In user assistance, using these words at the beginning of a sentence is *not* poor style. Using these words is much better than clumsy and overblown phrases such as *as a consequence of*, *as a result of*, *due to the fact that*, *owing to the fact that*, and so on.

However, bear in mind that most readers don't read instructions from start to finish. Thus, don't start a new *paragraph* with a word that relates to the previous paragraph. If you're doing so, this is an indicator that you probably shouldn't start a new paragraph here.

✖ No: *Due to the fact that you're reading this text, your manuals will improve.*

✔ Yes: *Because you're reading this text, your manuals will improve.*

Related rules

Blacklist: Overblown words 141

because / since / as 205

2.4.5   Feel free to end sentences simply

There's no rule against ending a sentence or phrase with a preposition, nor is there a rule that requires it.

- A sentence or phrase that ends with a preposition sounds less formal, but it might be harder to understand for readers who speak English as a second language.
- If the preposition is too far away from the verb, the sentence or phrase becomes hard to read.
- Only end sentences and phrases with a preposition if the sentence or phrase is short.

✖ No: Enter the text for which you want to search.

✔ Yes: Enter the text that you want to search for.

✔ Yes: Enter the text for which you want to search into the search field.

✔ Top: Enter the text that you want to search for into the search field.

✖ No: Enter the text that you want to search the document, the database, and the web for.

✔ Yes: Enter the text for which you want to search the document, the database, and the web.

✔ Yes: Enter the text that you want to search for in the document, in the database, and on the web.

2.4.6   Watch the order of words

The right order of words and phrases within a sentence can be a key factor in making your text easy to grasp and the instructions easy to follow.

1 The beginning and the end of a sentence are the most prominent positions. Place the key information at the beginning of the sentence. If that's not possible, try to place it at the end of the sentence (see *Always start with the main point* 19).

2 If you're describing an action, first state the goal, then state the action. Readers who don't want to reach the goal can then skip the rest of the sentence.

3 Follow the order that's determined by the natural succession of actions.

4 Arrange pairs and series from short to long and from simple to compound. In English, it's also common to arrange from the more specific to the more general; however, note that in some other languages this arrangement is vice versa.

5 Keep related words as close together as possible to prevent ambiguity and confusion.

1

✘ **No:**   *A memory card and some image editing software come with the camera.*

✔ **Yes:**   *The camera comes with a memory card and some image editing software.*

(The key component is the camera, so it goes to the most prominent position, which is the beginning of the sentence.)

2

✘ **No:**   *Use a pen to write a letter.*

✔ **Yes:**   *To write a letter, use a pen.*

3

✘ **No:**   *Click Save in the Options window.*

✔ **Yes:**   *In the Options window, click Save.*

(Readers must first find the **Options** window before they can find and click the **Save** button.)

✘ No: *The small, silver, rectangular cover plate is located on the bottom.*

✔ Yes: *The cover plate is located on the bottom. It's small, silver, and rectangular.*

(Readers must first know which object you're talking about before they can go looking for it.)

4

✘ No: *bicycles and cars*

✔ Yes: *cars and bicycles*

(The word *cars* is significantly shorter and simpler than *bicycles*, so it comes first.)

✘ No: *consumer electronics, software, and hardware*

✔ Yes: *software, hardware, and consumer electronics*

(*Consumer electronics* is the longest and the most general term, so it comes last. *Hardware* is related to *consumer electronics* because consumer electronics is also a type of hardware. Due to this close relation, *hardware* is positioned near *consumer electronics*. *Software* is related to *hardware* but not related to *consumer electronics*, so *software* goes to the opposite end of the list.)

5

✘ No: *He wrote three pages on how to use the computer for the user manual.*

✔ Yes: *He wrote three pages for the user manual on how to use the computer.*

(*User manual* and *pages* are related words, so it's better to place them near each other.)

Special rules for the English language

When you use **multiple adjectives** in one sentence, use this standard order:

1. Observation or opinion
2. Size and shape
3. Age
4. Color
5. Origin

6. Material

7. Qualifier (The qualifier is often an integral part of the noun.)

✘ No: *red, new, shiny car*

✔ Yes: **shiny, new, red car**

✔ Yes: **quality, 10-meter, round, used, silver, Swedish, stainless steel water pipe**

Follow the order **how – where – when**.

✘ No: *This morning he spoke well at the presentation.*

       (Unclear: Is this an exception? Doesn't he usually speak well?)

✔ Yes: **He spoke well at the presentation this morning.**

With verbs of movement, follow the order **place – manner – time**.

✘ No: *To the exhibition hall he went by taxi this morning.*

       (Unclear: Were there also some other places where he went with other means of transportation?)

✔ Yes: **He went to the exhibition hall by taxi this morning.**

Put **the more specific expressions before the more general ones**:

✔ Yes: **Monday, May 29, 2020**

✔ Yes: **Meet me at half-past seven tomorrow.**

✔ Yes: **We'll meet at 25 Freedom Square in Metropolis City at 9 a.m. local time on January 15, 2020.**

Related rules

Always start with the main point 19

Watch the position of modifiers 99

2.4.7 Watch the position of modifiers

Be aware of the fact that the position of modifiers such as *only*, *just*, *already*, *even*, *also*, *nearly*, *almost*, *merely*, or *always* may determine the emphasis or even the meaning of a sentence.

1 Place the modifier immediately *before* the word or phrase that you want to modify.

2 Exception: Only place the modifier between *to* and the rest of an infinitive if there's no other possibility.

1

✔ **Yes:** *We only work during office hours.*

(means: During office hours, we don't do anything else but work.)

✔ **Yes:** *We work only during office hours.*

(means: We don't work outside of office hours.)

✔ **Yes:** *Only we work during office hours.*

(means: We're the only ones who work during office hours.)

✘ **No:** *We almost worked seven years on the new product.*

(Implies: We almost worked, but mostly we just had a good time.)

✔ **Yes:** *We worked almost seven years on the new product.*

✘ **No:** *In the web shop, the list only shows the most popular items.*

(Implies: The list shows the items, but you can't order them here.)

✔ **Yes:** *In the web shop, the list shows only the most popular items.*

2

✘ **No:** *This is going to substantially decrease fuel consumption.*

(Here, *substantially* splits *to decrease*.)

✔ **Yes:** *This is going to decrease fuel consumption substantially.*

✘ **No:** *This is going to double fuel consumption nearly.*

(Here, it's not possible to place *nearly* after the noun.)

✔ **Yes:** *This is going to nearly double fuel consumption.*

Related rules
Be specific 23

2.4.8 Feel free to repeat a word

When it adds clarity, don't hesitate to use the same word over again.

Don't avoid using the same word twice in a sentence or in consecutive sentences because you think that this is poor style. In user assistance, the best style is clarity.

1 In sentences with *and* and *or*, make sure that it's clear which terms an attribute relates to.

2 Words such as *this, that, they, these, those, it, which*, and so on link ideas. Make sure that these pointers point unmistakably to one noun, phrase, or clause so that your sentence isn't ambiguous. If there's *any* possibility of confusion, repeat the noun.

3 Don't use awkward constructions like *the former ... the latter*. These constructions aren't ambiguous, but they're extremely hard to read. Most readers do remember the statements, but they don't remember which one came first. So they have to read the whole section again, which wastes their time, makes them feel stupid, and doesn't contribute to a positive user experience.

1

✘ **No:**   You need green paper and tape.

(Unclear: Must the tape also be green?)

✔ **Yes:**   You need green paper and some tape.

or:

You need green paper and green tape.

2

✘ **No:**   She told her colleague that her phone wasn't working.

(Ambiguous: Whose phone wasn't working? Her own phone or her colleague's phone?)

✔ **Yes:**   She told her colleague that the colleague's phone wasn't working.

or:

She told her colleague that her own phone wasn't working.

✗ No: *You can use the instrument to measure the PH value and the humidity of the soil. Note that this only works if the temperature is above 0 degrees Celsius.*

(Unclear: Do both measurements need a minimum temperature? Also unclear: Which temperature is important? The temperature of the soil? The air temperature? Maybe even both temperatures?)

✔ Yes: *You can use the instrument to measure the PH value and the humidity of the soil. Note that you can only measure the PH value if the soil temperature is above 0 degrees Celsius.*

or:

You can use the instrument to measure the PH value and the humidity of the soil. Note that you can only measure the humidity if the soil temperature is above 0 degrees Celsius.

or:

You can use the instrument to measure the PH value and the humidity of the soil. Note that both measurements only work if the soil temperature is above 0 degrees Celsius.

3

✗ No: *Type 657B devices are green. Type 657C devices are red. The former are made of plastics, whereas the latter are made of steel.*

✔ Yes: *Type 657B devices are green. Type 657C devices are red. Type 657B devices are made of plastics, whereas type 657C devices are made of steel.*

✔ Top: *Type 657B devices are green and made of plastics. Type 657C devices are red and made of steel.*

Related rules

Add syntactic cues [104]

Always use the same terms [123]

Be clear about what you're referring to [106]

Be parallel [32]

2.4.9   Use redundancies purposefully

In general, avoid including the same information twice. The shorter your documents are, the better.

Sometimes, though, if you don't repeat a certain piece of information, some users might miss it.

Redundancy is often especially important:

- in warnings
- if readers are inexperienced
- if it's likely that readers don't expect a certain kind of information
- when readers don't read the whole text

✗ No: *Warning:*
Use the washing machine only for washing clothes.

✔ Yes: *Warning:*
Use the washing machine only for washing clothes. Don't use the machine for washing your pet, or for washing objects such as shoes, dishes, or tools.

(Strictly speaking, it's obvious and logical that users mustn't use the machine for washing pets if you tell them that they can use the machine only for washing clothes. However, the word "only" is quite weak here. Some readers may not notice this little word, or they may not think about its implications. So it can be crucial to have some redundant information that explicitly tells users not to wash pets, shoes, dishes, and tools.)

✗ No: *Don't drink and drive.*

✔ Yes: *Warning:*
Don't drink and drive.

(The word *Warning* is redundant here because when you read the sentence, it's obvious that it's a warning. However, the redundant word *Warning* makes the statement a lot stronger.)

✗ No: *To apply for the job, please send us an email.*

(Although the statement is completely unambiguous, some applicants might mail a letter via the post office instead. Perhaps they believe that their chances are better if they send an application on paper because it's more difficult to ignore it or because it looks more presentable.)

✔ Yes: *To apply for the job, please send us an email. Please don't send any printed material.*

2.4.10   Add syntactic cues

Syntactic cues are words or punctuation marks that help readers to analyze the structure of a sentence more quickly and more reliably.

- Syntactic cues enhance readability.
- Syntactic cues can be especially helpful for readers who speak the document's language as a second language.
- Syntactic cues often eliminate ambiguities.

If you can add a syntactic cue, do so. The few extra characters or words are a good investment in clarity.

However, inserting a syntactic cue isn't always the best remedy for a poorly designed sentence. Sometimes it's better to rephrase a sentence completely or to make two sentences out of one (see *Make short sentences* 88).

The best way to get a good feeling for syntactic cues is to look at some examples:

✘ No: *Programs currently running are indicated by icons in the Task bar.*

✔ Yes: *Programs **that are** currently running are indicated by icons in the Task bar.*

✘ No: *The advanced search feature is especially helpful for users familiar with regular expressions.*

✔ Yes: *The advanced search feature is especially helpful for users **who are** familiar with regular expressions.*

✘ No: *You can print reports using the print utility.*

✔ Yes: *You can print reports **by** using the print utility.*

✘ No: *You can run macros using the Macro utility.*

(Ambiguous: Does the Macro utility run the macros or do the macros use the Macro utility?)

✔ Yes: *You can run macros **by** using the Macro utility.*

or

*You can run macros **that** use the Macro utility.*

✘ No: *The program continues processing data after restoring the database.*

✔ Yes: **The program continues to process data after it has restored the database.**

✘ No: *The operating system terminates the program if an error or exception occurs.*

✔ Yes: **The operating system terminates the program if an error or an exception occurs.**

✔ Top: **The operating system terminates the program if either an error or an exception occurs.**

✘ No: *If you choose option A and the program runs in auto-detection mode, something happens.*

✔ Yes: **If you choose option A and if the program runs in auto-detection mode, something happens.**

✘ No: *Your data must not include leading blanks and semicolons.*

(Ambiguous: Does the sentence mean only leading semicolons or all semicolons?)

✔ Yes: **Your data must not include semicolons and leading blanks.**

(Note: The syntactic cue here is the reversed order of *semicolons* and *blanks*.)

or:

Your data must not include leading blanks and leading semicolons.

✘ No: *The program was first published by company A and then modified by company B.*

✔ Yes: **The program was first published by company A and was then modified by company B.**

✘ No: *full- and part-time workers*

✔ Yes: **full-time and part-time workers**

✔ Top: **full-time workers and part-time workers**

Related rules

Feel free to repeat a word [101]

Use "and" with care [109]

Be clear about what you're referring to [106]

2.4.11 Be clear about what you're referring to

When using words like *this*, *these*, *that*, *those*, *it*, *they*, and *them*, make sure that it's clear which subject you're referring to.

If it avoids ambiguity or improves readability: Don't hesitate to repeat the subject as often as necessary (see *Feel free to repeat a word* 10↑).

Keep your text as stupid as possible. Usually, the topics that you're talking about are challenging enough.

✖ **No:** *The bomb is connected to a red and to a blue wire. Cut it to defuse the bomb.*

✔ **Yes:** *The bomb is connected to a red and to a blue wire. To defuse the bomb, cut the red wire.*

or:

The bomb is connected to a red and to a blue wire. To defuse the bomb, cut the blue wire.

✖ **No:** *DemoSoft can do A and B. This helps you to ....*

(Unclear: Does the word *this* relate to A, or does it relate to B, or does it relate to the combination of A and B?)

✔ **Yes:** *DemoSoft can do A and B. A helps you to ....*

or

DemoSoft can do A and B. B helps you to ....

or

DemoSoft can do A and B. Both help you to ....

✖ **No:** *The main problem that people run into with pronouns is not tying them to nouns.*

(Unclear: Who isn't tied to nouns: the people or the pronouns?)

✔ **Yes:** *The main problem that people run into with pronouns is not tying the pronouns to nouns.*

✔ **Yes:** *The main problem that a writer can run into with pronouns is not tying them to nouns.*

(Here you can use *them* because it can't relate to *writer*, which is singular.)

Related rules

Feel free to repeat a word 10↑

this / that / these / those 255

2.4.12   Use "then" with care

1 Don't use words such as *then* or *next* to describe a sequence of more than two steps. Use a numbered list instead (see *Writing procedures* 56).

2 If a sentence describes only two related steps, it's OK and often even necessary to link them with *and then* to emphasize the sequential nature of the two actions.

Put a comma before *and then* to separate both clauses clearly.

3 Don't use *then* to introduce a subordinate clause that follows an if-clause.

1

✘ No: To make a recording, first insert an empty disc into the DVD drive. Then select the channel that you want to record. Next, press the Start time key, and then enter the time when the recording should begin. Then, press the End time key and enter the time when the recording should stop. Finally, press the Record now key.

✔ Yes: To make a recording:

1. Insert an empty disc into the DVD drive.

2. Select the channel that you want to record.

3. Press the **Start time** key and enter the time when the recording should begin.

4. Press the **End time** key and enter the time when the recording should stop.

5. Press the **Record now** key.

2

✘ No: Insert the CD and double-click the folder symbol.

(Unclear: Is the order of steps important? Do you necessarily have to insert the CD first? Or can you equally click the folder symbol first?)

✔ Yes: Insert the CD, and then double-click the folder symbol that appears.

3

✘ No: If you follow this rule, then your documents are easy to read.

✔ Yes: If you follow this rule, your documents are easy to read.

2.4.13 Use "and" with care

Human language is looser than programming languages. In human language, the word *and* just tells the reader that there's *some* relation, but it doesn't tell anything about the nature of this relation. This often results in misinterpretation.

If the word *and* might be ambiguous, add a syntactic cue (see *Add syntactic cues* 104) or rephrase the sentence.

Often, you can easily split a long, complex sentence at the position of the word *and* (see *Make short sentences* 88).

✘ No: *The function returns "true" if the attribute "color" is "blue" and the attribute "size" is "small."*

(Unclear: Does this statement mean that *both* conditions must be true *at the same time* (Boolean logic IF color=blue AND size=small)? Or does it mean that only one condition must be true (Boolean logic IF color=blue OR size=small)?

✔ Yes: *The function returns "true" if the attribute "color" is "blue" and, at the same time, the attribute "size" is "small."*

or:

The function returns "true" when at least one of the following conditions is met:

* *attribute "color" is "blue"*
* *attribute "size" is "small"*

Related rules

Add syntactic cues 104

2.4.14 Don't use "and/or"

The phrase *and/or* is imprecise.

Either it's *and*, or it's *or*, but it can't be both. Don't burden the reader with the trouble of finding out what applies. It's your job to find out and then to communicate the facts clearly.

1 Replace the phrase *and/or* either with just *and* or with just *or*. If that's not possible, rephrase the sentence and explain the facts in more detail.

2 The same rules apply if *and/or* is implied in similar constructions with other words.

Note:
If you use *and/or*, there's no space character before and after the slash.

1

✘ **No:**   *You can archive and/or delete reports.*

✘ **No:**   *You can archive/delete reports.*

✔ **Yes:**   *You can archive and delete reports.*

or:

You can archive or delete reports.

or:

You can first archive and then delete reports.

2

✘ **No:**   *The procedure can read/write data.*

✔ **Yes:**   *The procedure can read and write data.*

✘ **No:**   *If the water is too hot/cold, adjust the temperature.*

✔ **Yes:**   *If the water is too hot or too cold, adjust the temperature.*

✘ **No:**   *If there's no input/output signal, ...*

✔ **Yes:**   *If there's neither an input signal nor an output signal, ...*

or:

If there's either no input signal or no output signal, ...

Related rules

Be specific 23

Use "and" with care 109

2.4.15 Don't use "(s)"

Don't create optional plurals.

Don't burden the reader with having to decide whether something should be singular or plural. This is your job.

Tip:
When in doubt, in most cases the plural is the correct choice because it covers both scenarios.

✖ No: *Wait for x minute(s).*
✖ No: *Wait for x minute/s.*
✔ Yes: *Wait for x minutes.*

✖ No: *The cable can be 1 to 10 meter(s) long.*
✔ Yes: *The cable can be 1 to 10 meters long.*

✖ No: *Here you can enter one/several user name/s.*
✔ Yes: *Here you can enter one or more user names.*

Related rules
Be specific 23

2.5 Writing words

Use words that are common, simple, and unambiguous.

Rules for writing on the word level

- *Use short, common words* 114
- *Watch for "…ed"* 116
- *Watch for "the … of" and for "of the"* 117
- *Watch for opening "It …" and "There …"* 118
- *Avoid abbreviations and acronyms* 119
- *Use technical terms carefully* 122
- *Always use the same terms* 123
- *Use contractions* 125
- *Avoid strings of nouns* 128
- *Avoid stacks of modifiers* 129
- *Avoid unnecessary qualification* 130
- *Use strong verbs* 132
- *Avoid gerunds that conceal a direct verb* 134
- *Avoid phrasal verbs* 135
- *Avoid idiomatic expressions* 136
- *Avoid buzzwords* 137
- *Avoid developer jargon* 138
- *Use fair language* 139
- *Blacklist: Overblown words* 141
- *Blacklist: Filler words* 159

Related rules

Writing in general 15
Writing topics 45
Writing sections 51
Writing sentences 87

2.5.1 Use short, common words

Don't show off your vocabulary. Reading infrequently used words requires more mental work than reading common words. This slows down readers and makes your text hard to understand.

Keep it simple and stupid (the KISS principle).

Use words that are common and short. Use uncommon and long words only if there is no simpler alternative.

Bear in mind readers who speak the document's language only as a second language.

Typical examples of words that you should replace with simpler alternatives are nouns with 3 or more syllables and foreign words.

Instead of the complex term ...	use the simple form ...
alphabetical character	*letter*
attempt	*try*
capability	*ability*
condition	*state*
determination	*choice*
employ	*use*
indicate	*show*, *tell*, *say*
indication	*sign*
inspect	*check*
location	*site*, *place*
modify	*change*
preserve	*keep*
require	*need*
terminate	*end*
transmit	*send*
utilization	*use*

For more examples, see *Blacklist: Overblown words* 141.

Related rules

Use technical terms carefully 122

2.5.2 Watch for "...ed"

Try to simplify or omit words that end with "...ed."

If "...ed" indicates passive voice, put the sentence into active voice (see *Use the active voice* 39).

✘ No: *Users who are located in France ....*
✔ Yes: **Users in France ....**

✘ No: *A study that was conducted by our company shows that ....*
✔ Yes: **A study by our company shows that ....**

✘ No: *centralized control*
✔ Yes: **central control**

✘ No: *improved results*
✔ Yes: **better results**

Related rules

▶ Use the active voice 39

2.5.3   Watch for "the ... of" and for "of the"

When possible, avoid the words *of* and *the*.

Exception: Possessives from company names, product names, feature names, and objects (see *Possessives* 247). However, in this case it's often better to use the name of the company, product, feature, or object as an adjective.

✘ **No:**   The purpose of this program is to print reports.
✔ **Yes:**   **This program prints reports.**

✘ **No:**   The assembly of computers is often done in China.
✔ **Yes:**   **Computers are often assembled in China.**

✘ **No:**   Some of the tasks require special knowledge.
✔ **Yes:**   **Some tasks require special knowledge.**

✘ **No:**   DemoSoft's key features
✔ **Yes:**   **the key features of DemoSoft**

✘ **No:**   Enter the disk's name.
✔ **Yes:**   **Enter the name of the disk.**
✔ **Top:**   **Enter the disk name.**

2.5.4 Watch for opening "It ..." and "There ..."

Review sentences that start with:

- "It is ...."
- "There is ...."
- "There are ...."

Often, you can find a more concise solution.

✘ No: *It is often the case that sentences are much too long.*
✔ Yes: *Often, sentences are too long.*

✘ No: *There are some functions that help you to save energy.*
✔ Yes: *Some functions help you to save energy.*

✘ No: *There's something wrong with this sentence.*
✔ Yes: *Something is wrong with this sentence.*

Related rules

Blacklist: Overblown words 14↑

2.5.5   Avoid abbreviations and acronyms

When in doubt, spell it out.

The advantage of improved clarity far outweighs the disadvantage of longer text. The use of too many abbreviations and acronyms is one of the most frequent reasons why readers fail to understand manuals and dislike reading them.

Abbreviate terms only if they appear in narrow table cells or other tight spaces.

Use acronyms only:

- if the acronym is also used in other materials that you can't change
- if the acronym is also used in the user interface of your product
- if you're sure that all readers know the meaning of the acronym
- if you need to use the term that the acronym stands for *very* often

If you can't avoid using an acronym

When you use an acronym, spell it out the first time, and then add the acronym in parentheses.

- In a printed manual, the "first time" is the page with the lowest page number.
- In online help, the "first time" is the topic that will be used either the earliest or the most frequently.

When spelling out the acronym, don't capitalize the words that make up the acronym unless the spelled-out form is a proper noun.

✔ **Yes:**   *original equipment manufacturer (OEM)*

✔ **Yes:**   *World Wide Web Consortium (W3C)*

If the pronunciation of an acronym isn't evident, provide a hint.

✔ **Yes:**   *WYSIWYG (pronounced "wiz-zee-wig")*

✔ **Yes:**   *W3C (W three C)*

Use capital letters without periods (exception: some geographic names).

✔ **Yes:**   *USB*

✔ **Yes:**   *XML*

✔ **Yes:**   *EU*

✔ **Yes:**   *U.S.*

✔ **Yes:**   *U.K.*

To form the plural of an acronym, add a lowercase *s* without an apostrophe.

✔ **Yes:**   *PCs*

✔ **Yes:**   *CPUs*

Use an apostrophe only if you need to form a possessive of an acronym.

✔ **Yes:**   *the OEM's products*

Don't include a generic term after an acronym if one of the acronym's letters stands for the same term.

✘ **No:**   *HTML language*

      (Note: the letter *L* already stands for the word *language* because HTML is the acronym for *Hypertext Markup Language*.)

✔ **Yes:**   *HTML*

✔ **Yes:**   *Hypertext Markup Language (HTML)*

Handling common abbreviations

In English, there are a number of common Latin abbreviations for frequently used word and phrases, such as "i.e." for "that is" or "e.g." for "for example."

Even though these Latin abbreviations are very common, avoid them when possible. Many readers don't know which words these terms abbreviate or they confuse the abbreviations. In particular, "i.e." and "e.g." are often confused.

Usually, the few extra characters needed to spell out the words are a good investment in clarity. Use Latin abbreviations only in situations where there isn't enough space to spell out the words, such as in narrow table columns.

Instead of ...	use ...
i.e.	*that is*, *in other words*
e.g.	*for example*
etc.	*and so on* Note: It's OK to use *etc.* for most audiences because it can't be confused with any other abbreviation.
et al.	*and others*

cf.	*compare*
viz.	*namely*
vs., v.	*versus, as opposed to* Note: It's acceptable to use *vs.* in headings.

2.5.6 Use technical terms carefully

Technical terms speed up communication between people who share the same expertise. For others, the same technical terms are just incomprehensible.

Use technical terms only:

- if you're writing for experts
- if a term is also used in other materials for the same user group as your document and you can't change these materials
- if a term is used in the user interface of your product

If you can't avoid using technical terms, explain them when using them for the first time.

- In a printed manual, the "first time" is the page with the lowest page number.
- In online help, the "first time" is the topic that will be used either the earliest or the most frequently.

✘ No: *This opens your default email client.*

(You don't need the technical term *client* here.)

✔ Yes: *This opens your default email program.*

✔ Yes: *Plug the cable into the USB port of your computer.*

(Here it's OK to use the technical term *USB port* because it's the only common name that exists for this type of interface. You can't change this name. It doesn't make any sense to invent a new name.)

Related rules

Use short, common words 114

Avoid abbreviations and acronyms 119

2.5.7 Always use the same terms

If you mean the same thing, use the same term.

If you use different terms, readers may think that you mean different things. Consistent terminology is also important when users want to skim your text for particular information, and when they use full-text search.

Variety is excellent for novels and for English tests at school. In technical writing, however, there's little room for creativity. Keep it simple and stupid (the KISS principle).

If you're writing sales texts and want to make your texts more conversational and vivid, vary the unimportant words rather than the important words.

Make a terminology list of the terms that you use and of the terms that you don't use (see *Be consistent* 29).

Don't use the terms that are on your blacklist within the visible text of your document, but add them as index keywords so that they appear in the alphabetical index. Readers who don't yet know which particular term you use can then find a topic even if they're looking for the "wrong" term.

Decision aids

When in doubt:

- use the word that's used in the user interface or printed on the product
- use the word that your users will be looking for when they skim a text
- use the simpler word
- use the shorter word

✖ **No:** *If you've purchased a new application, you must first install the software before you can use the program.*

✔ **Yes:** *If you've purchased a new program, you must first install this program before you can use it.*

Examples

Some typical examples of terms that need a decision on which words you use are:

- Do you say *computer*, or *PC*, or *machine*, or *client*, or *workstation*, or *unit*?

- Do you say *sound adapter*, or *sound card*?
- Do you say *pointer*, or *mouse pointer*, or *cursor*, or *mouse cursor*, or *arrow*, or *mouse arrow*, or *I-beam*?
- Do you say *keyboard shortcut*, or *quick access key*, or *shortcut key*, or *accelerator key*, or *hotkey*, or *speed key*, or *fast key*, or *quick key*, or *key combination*?

For recommendations, see *FAQ: Standard terms and phrases* 26 .

Related rules

Feel free to repeat a word 10

FAQ: Standard terms and phrases 26

Be consistent 29

2.5.8 Use contractions

Nowadays, it's perfectly OK to use contractions in all forms of user assistance.

- Contractions make your text more concise.
- Contractions are more conversational than the full forms.

However, don't use contractions:

- if you need to be formal, such as in legal texts (terms of use, end user license agreements, and so on)
- if you want to emphasize a word, in particular the word *not*
- in warnings (see *Writing warnings* 72)

✘ No: *If you have not used contractions before, do so now.*
✔ Yes: *If you haven't used contractions before, do so now.*

✘ No: *You mustn't use the washing machine to wash your pet.*
✔ Yes: *You must not use the washing machine to wash your pet.*

(There's an emphasis on the word *not*, so no contraction is used.)

List of common contractions

Full form	Contraction
are not	aren't
cannot	can't
can not	–
	Don't use *can't* as a contraction for *can not*. For the different uses of *cannot* and *can't*, see *can / may / might / must / should* 207.
could not	couldn't
did not	didn't
does not	doesn't
do not	don't
had not	hadn't
has not	hasn't

have not	haven't
he had he would	he'd
he will	he'll
I had I would	I'd
I will	I'll
is not	isn't
it is	it's Don't confuse with its.
it will	it'll
must not	mustn't (rarely used in American English)
need not	needn't
she had she would	she'd
she will	she'll
should not	shouldn't
that is	that's
there is	there's
they had they would	they'd
they will	they'll
they have	they've (rarely used in American English when not immediately followed by a verb)
was not	wasn't
we had we would	we'd
we will	we'll
we are	we're
were not	weren't
we have	we've
what is	what's
who are	who're

who have	who've
who is	who's
	(rarely used in technical documentation because it sounds very casual)
will not	won't
would not	wouldn't
you had	you'd
you would	
you will	you'll
you are	you're
you have	you've
	(rarely used in American English when not immediately followed by a verb)

Related rules

it's not / it isn't | 237 |

2.5.9   Avoid strings of nouns

Strings of nouns are hard to understand and sometimes even ambiguous.

Only use strings of nouns when they're the names of systems and you don't have the authority to simplify these names.

When a string of nouns is used as an adjective, use hyphens for clarification (see also *Hyphens* 183).

✖ No: *the device adapter card port signals*

(This is ambiguous. We can't tell whether it refers to "signals of the port" or to "port signals of the card.")

✔ Yes: *the device-adapter-card port signals*

✔ Top: *the port signals of the device adapter card*

✖ No: *technical documentation writing principles*

✔ Yes: *principles of writing technical documentation*

✔ Yes: *Local Area Network*

(You can't change this technical term.)

Related rules

Avoid stacks of modifiers 129

2.5.10   Avoid stacks of modifiers

When two or more modifiers appear before a noun, the meaning of the phrase often becomes ambiguous.

Using a hyphen in the right place can sometimes resolve the ambiguity. However, the phrase remains difficult to read and hard to understand.

For this reason: If there are two or more modifiers before a noun, always try to rephrase the sentence.

✖ **No:**   *Typical data conversion problem areas include ....*

✔ **Yes:**   *Typical data-conversion problem areas include ....*

✔ **Top:**   *Areas in which users typically have problems when converting data include ....*

✖ **No:**   *more effective methods*

✔ **Yes:**   *methods that are more effective*

or:

more methods that are effective

Related rules

Avoid strings of nouns 128

2.5.11   Avoid unnecessary qualification

Don't modify or qualify words that don't need to be modified or qualified.

Unnecessary qualification only adds empty calories, but it doesn't add any valuable information.

✘ No: *Both sentences tell you exactly the very same thing.*

✔ Yes: *Both sentences tell you the same.*

Additional examples:

Instead of ...	use the simple form ...
absolutely unique	*unique*
completely identical	*identical*
completely new	*new*
entirely complete	*complete*
exactly alike	*alike*
highly innovative	*innovative*
integral part	*part*
lift up	*lift*
new innovation	*innovation*
perfectly clear	*clear*
precisely the same	*the same*
repeat again	*repeat*
round circle	*circle*
the reason why	*the reason*
totally new	*new*
visible to the eye	*visible*

Related rules

Be concise 26

Blacklist: Overblown words 141

2.5.12   Use strong verbs

Use strong verbs that keep your texts clear, simple, and concise.

Many people tend to use "smothered verbs" because they feel that these verbs make their text sound more sophisticated. A smothered verb is a verb that is converted into a noun, which is then made the object of a less precise verb. (See the examples below.)

Avoid all sorts of smothered verbs.

✘ No: *We held a meeting and reached a decision on the improvement of our documents.*
✔ Yes: *We met and decided on how to improve our documents.*

✘ No: *The first step is the deletion of all needless words.*
✔ Yes: *The first step is to delete all needless words.*

✘ No: *Our software can be of help to you.*
✔ Yes: *Our software can help you.*

✘ No: *There are four screens within the wizard.*
✔ Yes: *The wizard consists of four screens.*

✘ No: *You can exert influence on the printing quality by using different types of paper.*
✔ Yes: *You can influence printing quality by using different types of paper.*

Additional examples:

Instead of ...	use the stronger, simpler verb ...
achieve reductions	reduce
conduct an analysis	analyze
conduct an investigation of	investigate
do an inspection of	inspect, check
form a plan	plan
give an answer to	answer
have knowledge of	know

have reservations about	doubt
have a concern	worry
hold a meeting	meet
make a decision	decide
make a distinction	distinguish
make a proposal	propose
make a recommendation	recommend
make a suggestion	suggest
provide a solution	solve
reach an agreement	agree

Related rules

Blacklist: Overblown words [141]

Use the active voice [39]

2.5.13 Avoid gerunds that conceal a direct verb

Generally, there's nothing wrong with using gerunds (verbs ending in -*ing*). However, avoid gerunds if they conceal a more direct verb.

✘ No: *Before being able to use the program, logging in is the first step that's required.*

✔ Yes: *Before you can use the program you must log in.*

✘ No: *Removing the cable should only be done when the device is turned off.*

✔ Yes: *Only remove the cable when the device is turned off.*

✔ Yes: *Removing the cable requires a special tool.*

✔ Yes: *Dropping the computer may destroy it.*

2.5.14   Avoid phrasal verbs

Readers who speak English as a second language often have difficulties understanding the exact meaning of phrasal verbs that don't have an equivalent in their first language.

If you're writing for an international audience, use one-word verbs when possible.

Typical examples:

Instead of ...	use ...
comes up	appears
fill out	complete
go back	return
go on	continue
look up	search, find
make sure	ensure
put off	postpone
put on	mount
put up with	tolerate
set up	arrange, configure, install
take off	remove
use up	consume

2.5.15 Avoid idiomatic expressions

Don't use idiomatic and colloquial expressions.

Idiomatic and colloquial expressions in another language are often hard to understand for readers who speak this language as a second language.

When you translate your document into foreign languages, idiomatic expressions can also cause a lot of trouble because often these expressions don't have any direct equivalents in other languages.

✘ No: We won't beat about the bush and tell you what our product can't do.

✔ Yes: We openly tell you what our product can't do.

✘ No: Setting up the device is as easy as pie and done in the twinkling of an eye.

✔ Yes: Setting up the device is easy and fast.

✔ Top: ...

(The best solution here is to leave out this statement completely. It doesn't add any significant value. It's actually a biased statement that may ruin your credibility if readers disagree (see Don't make judgments) 4↑.

2.5.16 Avoid buzzwords

Don't confuse user assistance with marketing materials.

In user assistance, use a trendy term or phrase (a buzzword) only if you'd use the same term or phrase even if it wasn't trendy.

If you're using materials from the marketing department as resources for your documents, pay especially close attention to eliminating needless buzzwords.

There are good reasons to avoid buzzwords even in promotional texts:

- While a buzzword is new, many people only have a vague understanding of its meaning.
- Most buzzwords go out of style very quickly. Documents that use buzzwords soon become outdated. Then, sometimes, they look old-fashioned; sometimes they look downright ridiculous.
- Many buzzwords don't add much to the meaning of a sentence (see *Blacklist: Overblown words* 14).

✘ **No:** *DemoSoft is a robust, cutting-edge tool with many user-friendly, powerful, and sophisticated features that will leverage your writing skills when writing user manuals.*

✔ **Yes:** *DemoSoft has many features that help you to write good user manuals.*

✘ **No:** *With its awesome 64 KB RAM and its cutting-edge tape drive, this innovative personal computer is ready for the coming millennium.*

(A few years after this sentence has been written, it makes readers smile.)

✔ **Yes:** *The computer has 64 KB RAM and can store data on a tape drive.*

(A few years after this sentence has been written, it's still OK.)

Related rules

Don't make judgments 4

Avoid developer jargon 135

2.5.17 Avoid developer jargon

Speak the language of the user. Only speak like a developer or programmer if you're speaking to developers or programmers.

Don't show off the fact that you belong to the elite group of those in the know.

If you're using some input from developers, such as technical specifications, pay especially close attention to eliminating all unnecessary technical terms and developer jargon.

✘ No: *application*
✔ Yes: *program*

✘ No: *CPU*
✔ Yes: *processor*

✘ No: *initiate*
✔ Yes: *start*

✘ No: *validate*
✔ Yes: *check*

Related rules

Avoid buzzwords [137]

Avoid abbreviations and acronyms [119]

Blacklist: Overblown words [141]

2.5.18   Use fair language

Be aware of the variety of people who use your product.

- Write in a gender-neutral way.
- Avoid cultural biases and stereotypes relating to religion, family structure, leisure activities, purchasing power, and a particular lifestyle.
- In examples, use both female and male first names. Use last names that reflect different cultural backgrounds.

Avoid using phrases like *he or she* or *his/her* in your attempt to be gender neutral. Phrases like these attract even more attention to gender and thus partly defeat your purpose. (Also, they bloat the text and slow readers down.)

To be gender neutral, the following often helps:

- Use a gender-neutral term instead of a gender-specific one.
- Don't speak of *users*. Address your readers directly as *you*.
- Put a sentence into the plural.

Note:
When using the plural to refer to a specific group of people, do so consistently throughout the whole document even if there's no problem with the gender in a particular case (see also *Be consistent* 29). Another advantage of the plural form is that it's often shorter because you don't need *a* or *the*.

✘ No: *If there's no sound in your earphones, ask the stewardess for assistance.*

✔ Yes: *If there's no sound in your earphones, ask the flight attendant for assistance.*

✘ No: *Each user must enter his password.*

✘ No: *Each user must enter her password.*

✘ No: *Each user must enter his/her password.*

✘ No: *Each user must enter their password.*

✔ Yes: *Each user must enter his or her password.*

✔ Top: *All users must enter their passwords.*

or:

Users must enter their passwords.

✘ No: *When a user wants to print a report, he or she needs to choose the Print command from the File menu.*

✔ Yes: ***When users want to print a report, they need to choose the Print command from the File menu.***

✔ Top: ***If you want to print a report, choose File > Print.***

or:

To print a report, choose File > Print.

Additional examples of gender-neutral terms:

Instead of ...	use the gender-neutral form ...
actress	actor
businessman	businessperson, professional
chairman	chair, chairperson
delegates and their wives	delegates and their spouses
fireman	firefighter
fisherman	fisher
foreman	supervisor
handyman	caretaker, repairer
hostess	host
housewife	homemaker
mailman	letter carrier
man hours	working hours, work hours
man-made	synthetic, artificial, hand-made, manufactured
manpower	workers, workforce, staff, employees, personnel
policeman	police officer
repairman	repairer, technician
salesman, salesgirl, saleslady	salesperson, representative
stewardess	flight attendant
tradesman	tradesperson
waitress	waiter, server
watchman	security guard
workmen	workers

2.5.19 Blacklist: Overblown words

Many of the given writing rules encourage you to use common and simple words. The following list is a collection of frequently used words and phrases that you can easily replace with more common or shorter alternatives.

You can use the list in two ways:

- You can browse the list to get a better feeling of how to write plain English.
- You can use the list to look up a better alternative for a particular term.

Instead of the complex term ...	use the simple term ...
A	
a great deal of	much
a large number of	many
a little less than	almost
a number of	some, several, many
a percentage of	some
a small number of	a few
a sufficient number	enough
absolutely essential	essential
absolutely sure	sure
absolutely unique	unique
accelerate	speed up
accomplish	do, finish, carry out
accordingly	so
acquire	buy, get
activate	start
active participation	participation
actual experience	experience
add an additional	add
additional	extra, more
adequate number of	enough

adjacent	*next to*
administer	*manage, control*
admissible	*allowed, acceptable*
advance planning	*planning*
advantageous	*useful, helpful*
adverse	*harmful*
all of the	*all (the)*
alphabetic character	*letter*
alteration	*change*
alternative	*choice, other*
alternatively, you can also ...	*alternatively, you can ...*
and also	*and, also*
anticipate	*expect*
annually	*yearly*
apparent	*obvious, clear, plain*
appear	*seem*
application	*use, program*
appropriate	*proper, right, suitable*
approximately	*about, roughly*
as a consequence	*consequently*
as a consequence of	*because*
as a general rule	*generally, usually*
as a matter of fact	*in fact* (or omit)
as of the date of	*from*
ascend	*rise*
assist	*help*
assistance	*help*
at a later date	*later*
at all times	*always*
at an early date	*soon*
at present	*now*
at regular time intervals	*regularly*

at that point in time at that moment in time	*then*
at this point in time at this moment in time	*now*
at the moment	*now*
at the place	*where*
at the present time	*now*
attempt	*try*
attributable to	*due to, because of*
authority	*right, power*
authorize	*allow, let, approve*
autonomous	*independent, free*
B	
based on the fact that	*due to, because*
basic fundamentals	*fundamentals*
because of the fact that	*because*
beneficial	*helpful, useful, good for*
benefit	*help*
be able to	*can*
be in a position to	*can*
be in possession of	*have*
be of importance	*be important*
be of the opinion	*think, believe*
be present	*be at, be there*
be unable to	*can't*
blend together	*blend*
by means of	*by*
by the use of	*by, with*
C	
capability	*ability*
capacity	*ability, size, space*
category	*kind, class, group*
cease	*end, finish, stop*

cf.	*compare*
clarification	*explanation*, *help*
close proximity	*proximity*, *near*
come to a stop	*stop*
come to an end	*end*
commence	*start*, *begin*
communicate	*tell*, *write*, *talk*, *telephone*
compile	*make*, *collect*
completely identical	*identical*
completely new	*new*
completely surrounded	*surrounded*
comply with	*keep to*, *meet*, *follow*
component	*part*
comprise	*make up*, *include*
compulsory; it is compulsory	*you must*
conceal	*hide*
conceive	*imagine*, *think up*
concerning	*about*, *on*, *for*
condition	*state*
conduct	*carry on*, *do*, *run*
conduct an analysis	*analyze*
consequence	*result*
consequently	*so*
considerable	*many*, *much*, *great*, *important*
constitute	*make up*, *form*
construct	*build*, *make*
consult	*ask*, *talk to*, *check with*, *meet*, *read*
contain	*have*
convenient	*handy*, *useful*
conversion	*change*
cumulative	*added up*
current status	*status*

currently	(right) now
curtail	limit, reduce, shorten
customary practice	practice
D	
de facto	actual, real
deactivate	close, shut off, stop
decrease	drop, cut back
deduct	take away from, take off
defective	broken, faulty
defer	put off, delay
deficiency in / of	lack of
demonstrate	show, prove
depict	show
descend	fall
desire	wish, want
despite the fact that	though, although
detailed information	details
detect	find
determine	decide, set
diagnose	find out
difficulties	problems
diminish	lessen, reduce, decrease
disclose	tell, show
disconnect	cut off, unplug
discontinue	stop, end, give up
discrete	separate
discuss	talk about
disintegrate	break up, fall apart
display	show
distribute	spread, hand out
do an inspection of	inspect
dominant	main

due to the fact that	*because, as, due to*
duplicate	*copy, repeat*
during the course of	*during*
during the time that	*while*
during which time	*while*
E	
e.g.	*for example*
each one each and every one	*each*
elevate	*lift, raise*
eligible	*qualified, allowed*
eliminate	*remove, get rid of*
elucidate	*explain*
emphasize	*stress*
employ	*use*
empower	*allow, let*
encounter	*come upon*
end result	*result*
endeavor	*try*
entirely complete	*complete*
envisage	*foresee, imagine, see*
equilibrium	*balance*
equivalent	*equal, the same*
erroneous	*wrong, false*
establish	*show, find out, set up*
etc.	*and so on, and so forth*
evaluate	*test, check*
evident	*obvious, plain*
evince	*show*
exactly alike	*alike*
excessive number of	*too many*
exclusively	*only*
execute	*run*

exhibit	show
expend	spend
expire	end
explicit	plain, clear, exact, precise
extend	lengthen, offer, reach, stretch
extensive	large, broad, wide
exterminate	destroy, wipe out
F	
facilitate	ease, help, lighten, make easy
factor	reason, cause
fail to	don't
feasible	possible, can be done
few in number	few
final result	result
finalize	complete, finish, end
following	after
for a period of	for
for the duration of	during, while
for the purpose of	to, for
for the reason that	because
for example ... and so on	for example
forthwith	immediately
fortunate	happy, lucky
forward	send
fracture	break
frequently	often
from time to time	occasionally
fully compatible	compatible
function	act, work
functionality	features
fundamental	basic, real
furnish	give, send

furthermore	then, also, and
future plans	plans
G	
general public	public
generate	produce, make, give, create
give an answer to	answer
give authorization	authorize
give due consideration	consider, think
gives you the opportunity to	lets you, gives you the chance to
grant	give
H	
has a negative impact	harms, hurts
have a preference for	prefer
have knowledge of	know
hazardous	dangerous, risky, unsafe
hence	so
hereafter	after this takes effect
herein	here
hereof	of this
highly innovative	innovative
hitherto	until now, so far
I	
i.e.	that is
identical	the same
if at all possible	if possible
if that were the case	if so
illuminate	light up, make clear
illustrate	show, explain, draw
illustration	picture
immediately	at once, now
immovable	firm, fixed, set
imperative	pressing, urgent, vital

imperceptible	*hard to see, hidden, slight*
implement	*do, carry out, program, build in, set up*
imply	*suggest, hint at*
in a case in which	*when, where*
in a way so that	*so that*
in addition (to)	*and, also, as well as*
in advance of	*before*
in all cases	*always*
in case	*if*
in close proximity	*near, close*
in lieu of	*instead of, in place of*
in many cases	*often*
in most cases	*usually, mostly*
in order that	*so that*
in order to	*to, for*
in other words	*or, that is*
in reference to in regard to in relation to in respect to	*about, concerning*
in some cases	*sometimes*
in the absence of	*without*
in the course of	*during, while*
in the event of/that	*if*
in the majority of cases in the majority of instances	*usually, mostly, generally*
in the near future in the not too distant future	*soon*
in the proximity of	*about, near*
in the vicinity of	*near, close to, around*
in view of the fact that	*because*
incorrect	*wrong*
increase	*gain, go up, grow, rise*
indicate	*show, suggest, say, tell, mean*

indication	*sign*
indispensable	*essential*, *much-needed*, *vital*
individual (noun)	*person*
inform	*tell*
initial	*first*
initially	*at first*
initiate	*begin*, *start*
input	*comments*, *opinion*
inquire	*ask*
insert	*put in*
inside of	*inside*
inspect	*check*
instances	*cases*
instrument	*means*, *tool*
insufficient	*not enough*
integral part	*part*
intention	*aim*, *goal*, *plan*
invoke	*start*, *run*
irregardless of irrespective of	*regardless of*
is able to	*can*
is provided with	*has*
is unable to	*cannot*
issue	*give*, *send*
it is apparent that	*apparently* (better: omit)
it is clear that	*clearly* (better: omit)
it is evident that	*evidently* (better: omit)
it is likely that	*likely*
it is mandatory to	*you must*
it is obligatory to	*you must*
it is obvious that	*obviously* (better: omit)
it is often the case that	*often*
it is probable that	*probably*

it will be necessary to	you will need to
K	
keep an eye on	watch
L	
large number of	many
last of all	last
left hand left-hand side	left
lengthy	long
leverage	take advantage of, use
lift up	lift
locate	find
location	place, spot
M	
magnitude	extent, size
major	main
major portion of	most of
majority of	most
make a decision	decide
make a distinction	distinguish
make a payment	pay
make a recommendation	recommend
make a suggestion	suggest
manifest	show
manner	way
manufacture	make, produce, build
marginal	small, slight
maximum	most, largest, longest, greatest
might possibly	might
minimal	least, small, smallest
minimum	least, smallest
mix together	mix

modification	*change*
modify	*change*
monitor	*check*, *watch*
month of June	*June*
N	
navigate to	*go to*
never at any time	*never*
nevertheless	*but*, *however*, *even so*
new innovation	*innovation*
notify	*tell*, *let ... know*
notification	*notice*
notwithstanding notwithstanding the fact that	*even if*, *despite*, *still*, *yet*, *although*, *even though*
numerous	*many*
O	
objective (noun)	*aim*, *goal*
observe	*see*, *watch*
obsolete	*out-of-date*
obtain	*get*, *gain*, *receive*
obvious	*clear*, *plain*
occur	*happen*, *take place*
of a technical nature	*technical*
offer	*give*
on a daily basis	*daily*
on a few occasions	*occasionally*, *sometimes*
on a number of occasions	*often*
on account of	*because*
on an ongoing basis	*continually*
on behalf of	*for*
one and the same	*the same*
on numerous occasions	*often*
on the grounds that	*because*
operate	*run*, *work*

optimum	best, ideal, most, greatest
option	choice
or, alternatively	or
other alternatives	alternatives
outcome	result
owing to the fact that	because
P	
paramount	main, chief
partially	partly
participate	join in, take part
particulars	details
past experience	experience
per annum	a year
per day	a day
perfectly clear	clear
perform	do, carry out, make
period of time	period, time
permanent	lasting
permissible	allowed
permit	allow, let
personal opinion	opinion
philosophy	idea, system, view
place	put
portion	part, piece, share
possess	have, own
potential hazard	hazard
practically	almost, nearly
pragmatic	practical
precisely the same	the same
prerecorded	recorded
preserve	keep, protect
previous, previously	last, before, earlier

principal	*main*
prior	*earlier*
prior to	*before*
prioritize	*rank*
probability	*chance, likelihood*
proceed	*go, go ahead, start*
procure	*get*
proficient	*skilled*
prohibited	*forbidden*
projected	*estimated*
proportion	*part*
provide	*give, offer, send, supply*
provided that	*if, as long as*
provisions	*terms, rules*
provoke	*cause*
proximity	*nearness, closeness*
purchase	*buy*
R	
rationale	*reason, thinking*
reach an agreement	*agree*
really important	*important*
reconsider	*look again at, think again about*
reduce	*cut*
reduction	*cut*
refer (back) to	*see*
referred to as	*called*
regarding	*on, about*
register	*sign up*
regulate	*control*
relating to	*about, on*
remain	*stay*
remainder	*rest, others*

repeat again	repeat
represent	be, stand for, show
request	ask for
require	need
requirement	need, demand, wish
reservation	doubt
reside	live, be
respecting	about
respond	answer, reply
restriction	limit, limitation
retain	keep
return again	return
reveal	show, uncover
review	check, go over
right right-hand side	right

S

segment	part
seldom if ever	rarely
send an invitation to	invite
shall	will
significant	big, important, serious
so as to	to
sole, solely	only
sophisticated	complex
stakeholder	(say which person or which group of persons you mean)
start off with	start with
state	say
status	state
stringent	strict, tight
subject matter	subject
subsequent	next, following, later

subsequent to	*after, later*
subsequently	*later*
substantial	*big, large, great, significant*
such as ... and so on	*such as*
such as ... for example	*such as ..., for example ...*
suffice	*be enough*
sufficient sufficient number of	*enough*
supplementary	*extra, more*
supply	*give, send*
T	
take a look at ...	*look at ...*
terminate	*stop, end*
termination	*end*
that is to say	*that is*
the fact that	*that*
the majority of	*most*
the manner in which	*how*
the only difference being that	*except that*
the question as to whether	*whether*
the reason is because	*the reason is*
the very same thing	*the same*
therefore	*so*
through the use of	*with, by*
time period	*time*
to date	*so far, up to now*
totally new	*new*
transfer	*change, move*
transform	*alter, change*
transformation	*change*
transmit	*send*
U	
ultimate	*final, last*

ultimately	in the end, finally
unavailability	lack of
under the provisions of	under
unoccupied	free, empty
utilization	use
utilize	use (utilize is OK when you mean "to find a practical use for")
V	
validate	check
velocity	speed
verify	check, find out if
very final	final
very precise	precise
via	by, with, through, by means of
viable	possible
virtually	almost
visible to the eye	visible
voluminous	bulky, big
W	
ways and means	ways
whatsoever	whatever, what, any
whether or not	whether
with regard to	about, concerning, for
with respect to	for, on, about
with the exception of	except
with the result that	so that
Y	
you're requested to	you must

Related rules

Blacklist: Filler words 159
Be concise 26

2.5.20   Blacklist: Filler words

When you encounter a filler word or phrase, check whether you can omit it. This makes your text shorter and improves its readability.

Many of the words and phrases listed below indicate a contrast or consequence. At first glance, such words may seem to be important, but often you can just omit them without losing any information.

Also, filler words are mostly vague or judgmental, which alone is a good reason to get rid of them. Omit the filler word or phrase completely, or replace it with a more precise term.

For example, consider omitting or replacing:

- *a total of*
- *absolutely*
- *actually*
- *after all*
- *altogether*
- *all but*
- *already*
- *also*
- *always*
- *any*
- *anyway*
- *as a matter of fact*
- *as such*
- *basically*
- *certainly*
- *definitely*
- *different*
- *even*
- *ever*
- *existing*
- *however*
- *in fact*
- *indeed*
- *it should be understood*

- *just*
- *more or less*
- *mostly*
- *obviously*
- *of course*
- *perfectly*
- *quite*
- *rather*
- *really*
- *simply*
- *so*
- *some*
- *somewhat*
- *still*
- *sure, surely*
- *that is to say*
- *though*
- *totally*
- *truly*
- *very*
- *virtually*
- *well, ...*
- *yet*

Related rules

Blacklist: Overblown words [141]

Be specific [23]

Be concise [26]

Avoid unnecessary qualification [130]

2.6 FAQ: Spelling and punctuation

Even small errors undermine your credibility.

Don't underestimate the importance of correct spelling and punctuation. Flawless instructions are much more trustworthy than those with spelling mistakes and typos.

> **Important:** Electronic spelling checkers can identify many mistakes, but they can't find them all. Electronic spelling checkers, grammar checkers, and other language tools don't work as a substitute for editing and proofreading by a human.

FAQ

When writing a text, you probably don't want to waste your time browsing bulky grammar reference manuals and style guides. For this reason, we've compiled quick answers to the most frequent questions that arise when writing instructions and technical documents.

Tip:
The listed topics include not only frequently asked questions but also frequently made mistakes. Even if you don't have any particular question now, take the time to skim the topics in this section for details that you might not yet be aware of.

2.6.1 Apostrophes

1 Use apostrophes to form the possessive case of nouns.

- It's OK to form possessives from acronyms.
- Avoid the possessive forms of company names, product names, and feature names.
- If a singular noun ends with an s, x, or z, add an apostrophe and an s as with any other noun.
- With plural nouns that end with an s, put the apostrophe after the s.
- With plural nouns that *don't* end with an s, add an apostrophe and an s as with singular nouns.

2 Use apostrophes to indicate missing letters or missing numerals in a contraction.

Differentiate between the contraction *it's* (it is) and the possessive pronoun *its*.

3 Don't use an apostrophe to indicate the plural of a singular noun, the plural of an acronym, or the plural of a numeral.

Only use apostrophes to form the plural of single letters, symbols, and mathematical signs.

1

✔ **Yes:** *the car's engine*
✔ **Yes:** *an OEM's product*

✘ **No:** *Demo Corporation's headquarters*
✔ **Yes:** *the headquarters of Demo Corporation*

✔ **Yes:** *the box's contents*
✔ **Yes:** *James's car*
✔ **Yes:** *the boss's desk*

✔ **Yes:** *the lady's dress*
✔ **Yes:** *the ladies' dresses*
✔ **Yes:** *the Joneses' house*
✔ **Yes:** *children's requirements*

2

✔ **Yes:** *Don't drink and drive.*

✔ **Yes:** *It's important to make backup copies of your files.* (contraction)

but: *The battery loses its power.* (possessive pronoun)

✔ **Yes:** *in the early '60s.*

3

✔ **Yes:** *Computer programs make writing easy.*

but: *We recommend reading the program's manual.* (possessive)

✔ **Yes:** *The product is distributed by 10 OEMs.*

✔ **Yes:** *The technology was developed in the early 1980s.*

✔ **Yes:** *Type three 5s in a row.*

✔ **Yes:** *Make sure you dot your i's.*

✔ **Yes:** *The program replaces unrecognizable characters with *'s.*

✔ **Top:** *The program replaces unrecognizable characters with asterisks.*

Typographical conventions

With the invention of the typewriter, a neutral, straight form of the apostrophe was created so writers could use the same typewriter key for apostrophes as well as for opening and closing single quotation marks. This form is known as the typewriter apostrophe, and it is also available on standard computer keyboards. In contrast to the typewriter apostrophe, the typographic apostrophe looks like the numeral "9."

To give your document a truly professional appearance, use typographic apostrophes rather than typewriter apostrophes.

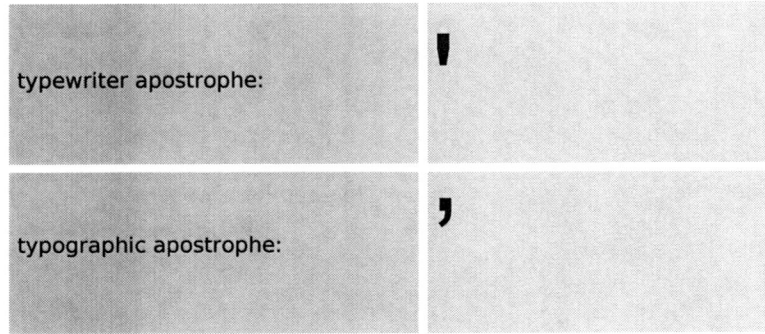

typewriter apostrophe:

typographic apostrophe:

How to enter typographic apostrophes

On standard computer keyboards, there's a key for the typewriter apostrophe, but there's no key for the typographic apostrophe.

Nowadays, many authoring tools automatically insert a typographic apostrophe when you type a typewriter apostrophe.

Tip:
If your authoring tool doesn't support typographic apostrophes, you can also use the Special Characters Script included in the indoition Hotkey Script Collection for Writers and Translators to enter them (for more information, visit *http://www.indoition.com*).

To enter a typographic apostrophe manually:

- On a computer that runs Windows: Hold down Alt and type 0146 on the numeric keypad.
- On a computer that runs Mac OS: Press Alt+Shift+].
- On a computer that runs Linux: Press Ctrl+Shift+U, then type 2019, then press Enter.

The HTML code for the typographic apostrophe is `&rsquo;`.

The Unicode code point for the typographic apostrophe is U+2019.

Related rules

Possessives 247

2.6.2 Capitalization of headings

There are two alternative styles for headings and subheadings:

- **Caps and lowercase style**
  Example: *This Is a Caps and Lowercase Heading*

- **Sentence style**
  Example: *This is a sentence-style heading*

If you have a company style guide that dictates capitalizing headings, follow this style. If you're free to choose, use sentence-style capitalization, which is more modern and easier to use.

Make sure that your headings are formatted so that they can be clearly identified as headings.

Apply the same rules for figure titles and table titles (if you have any). However, always use sentence-style capitalization for figure callouts, for texts within figures, for table row headings, and for table column headings.

Advantages of caps- and lowercase-style capitalization

- Caps- and lowercase-style capitalization is more widely used (exceptions: sales literature, newspaper articles, and web articles).

- Visually, headings can be identified more easily, especially if the format of the headings resembles the format of body text.

Advantages of sentence-style capitalization

- Sentence-style capitalization is easier to read, especially on a screen and within a long table of contents.

- You don't have to follow any special conventions, so you're less prone to making errors.

Mixing capitalization styles

As an option, you can mix capitalization styles to highlight different hierarchical levels. For example, you can use caps- and lowercase-style capitalization for level 1 headings but sentence-style capitalization for all levels below. Or you can use caps- and lowercase-style capitalization for topic headings and sentence-style capitalization for subheadings.

Basic rules for caps- and lowercase-style capitalization

If you use caps and lowercase style:

- Capitalize the first word and the last word, no matter what these words are.
- Capitalize the first word after a colon, no matter what this word is.
- In general, capitalize all other words as well unless they're mentioned in the following rules.
- Don't alter the capitalization of abbreviations, feature names, product names, company names, and case-sensitive words. If one of these words is the first word, try to rewrite the heading.
- Don't capitalize articles (*a*, *an*, *the*).
- Don't capitalize coordinate conjunctions (*and*, *as*, *but*, *or*, *nor*, *so*, *for*, *yet*).
- Capitalize *Is*, *Are*, *Be*, *If*, *It*, *Its*, *That*, *This*, *Than*.
- Don't capitalize the following prepositions and adverbs: *at*, *by*, *for*, *of*, *in*, *up*, *on*, *to*. Many style guides also suggest not capitalizing *from*, *into*, *off*, *onto*, *out*, *over*, and *with*, *without*, *within*.
  However, always capitalize the mentioned prepositions if they're part of a verb phrase, such as *Set ... Up*, *Plug ... In*, and so on.
- Capitalize each word of a hyphenated compound if the word after the hyphen is a noun or proper adjective, or if the words have equal weight. Don't capitalize a second word that modifies the first word, such as *Follow-up*, *Add-in*.
- If a short word that would usually be lowercased according to listed rules is used in parallel with a capitalized word of equal significance, capitalize the short word as well.
- In words that use the letter "e" as a short form for "electronic," capitalize both the "E" and the first letter.

✔ **Yes:**   *How to Install the Software*

✔ **Yes:**   *Setting Up the Computer*

✔ **Yes:**   *What Is a Device Driver?*

✔ **Yes:**   *To Use This Manual*

✔ **Yes:**   *How to Use the Manual*

✔ **Yes:**   *How You Can Use the Product*

✔ **Yes:**   *A How-To Approach*

✔ **Yes:**   *Using the 64-bit Version*

✔ **Yes:**   *Risk of Accidents In and Around Airports*
          (Note: *In* and *Around* are used in parallel, so *In* is written with a capital letter as well.)

✔ **Yes:**   *E-Learning and Multimedia*

2.6.3 Capitalization of products and features

For product names and feature names, follow the same capitalization style that is:

- registered as a trademark
- used on the product and on the product's packaging
- used in other written materials, such as brochures, web sites, and so on
- used for other products from the same company

When in doubt, don't capitalize.

Don't capitalize additional words such as *card*, *driver*, and so on if these words aren't an integral part of the registered product name.

Capitalize industry-standard terms only if the rest of the industry does so.

✔ Yes: DemoCorp laser printers

✔ Yes: DemoCorp Smart Ultra Superb laser printer

✔ Yes: In addition to conventional laser printing, the new Ultra Superb laser printing technology enables you to ....

Handling the capitalization of product names and function names in headings

Don't alter the capitalization of product names and function names in headings. If a product name or function name starts with a lowercase letter, avoid beginning a heading with this word.

See also Capitalization of headings 165

Handling lowercase letters at the beginning of a sentence

If a product name or function name starts with a lowercase letter, don't capitalize this letter at the beginning of a sentence. If possible, rephrase the sentence.

✘ No: calc(x) is one of the most important functions.

✔ Yes: One of the most important functions is calc(x).

✔ Top: The calc(x) function is one of the most important functions.
(This is the best alternative because the key information is at the beginning of the sentence.)

2.6.4 Colons

The colon creates anticipation. It tells the reader that the information is still incomplete. Don't hesitate to use a colon where it serves this purpose. What stands before the colon doesn't have to be a complete sentence.

Use colons to:

- introduce a list
- introduce an example
- introduce an explanation or elaboration of what preceded the colon
- separate two clauses when the second clause explains the first

Don't use colons *at the end* of headings and subheadings. Here, the formatting already does the job of creating anticipation, so you don't need the colon. However, it sometimes makes sense to use a colon *within* a heading. If you must use a colon within a heading, capitalize the first word after the colon.

Place colons outside quotation marks and parentheses (international style).

✔ **Yes:**   *Creating a manual consists of three main tasks: structuring, designing, and writing.*

✔ **Top:**   *Creating a manual consists of three main tasks:*

- *structuring*
- *designing*
- *writing*

✘ **No:**   heading: *How to Write a Manual*

✔ **Yes:**   heading: *How to Write a Manual*

✔ **Yes:**   heading: *Setting Up: A Beginner's Guide*

Capitalization

Only capitalize the first word after the colon:

- if the word starts a full sentence
- if the word is a proper noun
- in headings

Related rules

2.6.5   Commas

Commas divide a sentence into its components, which are easier to process for the human brain than the complete sentence. A comma is a working aid for the reader. So when a comma is optional, go ahead and use it. It simplifies your text and prevents misinterpretation.

Don't add a comma just because you "feel" that there should be one. Commas can only guide readers reliably if comma placement follows widely recognized conventions.

The following sections sum up the key rules as briefly as possible. The given examples cover all major scenarios that typically occur in technical writing.

Commas separate list items

Insert a comma before the final *and* and before the final *or* in a list. The comma is optional here, but it's helpful.

✔ **Yes:**   *Before baking a cake, set aside some flour, eggs, sugar and salt.*

✔ **Top:**   *Before baking a cake, set aside some flour, eggs, sugar, and salt.*

Sometimes a comma changes the meaning of a sentence. If there's any risk that readers might confuse the facts, use parentheses or restructure the sentence.

✖ **No:**   *The meeting was organized by John, my boss, and me.*

(Unclear: Was the meeting organized by three people—John, my boss, and me? Or was it organized by two people—my boss, whose name is John, and me?)

✔ **Yes:**   *The meeting was organized by John (my boss) and me.*

or:

The meeting was organized by John, by my boss, and by me.

Commas separate adjectives only if each adjective modifies the noun

Use a comma to separate two adjectives if each adjective modifies the noun alone.

✔ **Yes:**   *a long, heavy pipe*

Don't use a comma if the second adjective is essential to the meaning of the phrase.

✔ **Yes:** *the local heavy industry*

Commas separate main clauses

Use commas to separate main clauses, especially those with *but, and*, and *or*. Commas are optional here, but helpful.

Usually there's no comma when the subject of the second clause is left out because then the second clause is no longer a full-fledged main clause.

✔ **Yes:** *The manual is excellent, but a few things are missing.*

✔ **Yes:** *Setup analyzes your system, and then it copies the required files to your hard disk.*

but: *Setup analyzes your system and then copies the required files to your hard disk.*

Don't be fooled by subordinating conjunctions. Subordinating conjunctions always join a *subordinate* clause to a main clause. For this reason, if you have a subordinating conjunction, the second part of the sentence is NOT a main clause but always a *subordinate* clause. So there is NO comma.

The most common subordinating conjunctions are:

- after
- although
- as, as long as, as much as, as soon as
- as if, as though
- because
- before
- even if, even though
- how
- if
- now that
- provided (that)
- since
- so that
- that
- though
- till, until
- unless

- when, whenever
- where (unless "where" is used in a non-defining relative clause where there must be a comma)
- wherever
- while

✔ Yes: *The indicator light flashes because the door is still open.*

(but: Because the door is still open, the indicator light flashes.)

✔ Yes: *The measurement starts as soon as you press the* **Test** *switch.*

(but: As soon as you press the **Test** switch, the measurement starts.)

In some rare cases, when a verb is in the negative, you may need to put in a comma or to rephrase a sentence to avoid confusion.

For example, the sentence "The engine did not start because the temperature was too cold." could mean two things:

- The engine did not start. The reason was that the temperature was too cold.
- The engine started, but the reason why it started wasn't the cold temperature.

Commas separate important ideas from less important ideas

Use commas when you insert an extra comment into the middle of a sentence.

Tip:
When you need to separate ideas with commas, this often indicates that you should split the sentence into two separate sentences or that you should rephrase the sentence.

✔ Yes: *This command, which can also be used for printing, is only available in the Pro version.*

✔ Top: *This command is only available in the Pro version. You can also use it for printing.*

Commas separate a subordinate clause only if the subordinate clause stands before the main clause

Use commas to separate subordinate clauses only if the subordinate clause stands *before* the main clause.

✔ Yes: *Choose Save from the File menu to save the file.*

✔ **Yes:**   *To save the file, choose* **File** > **Save**.

(Note: This version is better because it states the goal before the action.)

Commas separate introductory phrases

Commas are optional but helpful after an introductory phrase.

✔ **Yes:**   *On a sunny day like this, wear sunglasses.*

✔ **Yes:**   *However, don't wear dark sunglasses when driving a car.*

✔ **Yes:**   *After taking off your sunglasses, put them back into the case.*

✔ **Yes:**   *For details, read chapter ABC.*

Don't be fooled by a main sentence that looks like an introductory phrase.

✔ **Yes:**   *When the indicator light flashes, insert new paper.*

✔ **Yes:**   *Insert new paper when the indicator light flashes.*

("Insert new paper" at the beginning of the sentence isn't an introductory phrase here, but the main clause. So there's no comma.)

When a sentence begins with "Also" or "So," it's not always easy to decide whether the words "Also" or "So" are an introductory phrase.

If you can replace "Also" with "In addition," it's usually an introductory phrase and thus followed by a comma. If "Also" closely relates to a statement in the previous sentence, it's not an introductory phrase and there is no comma.

If the word "So" at the beginning of a sentence is essentially just a filler word that could be omitted, it's an introductory phrase and thus followed by a comma. If the word "So" at the beginning of a sentence suggests some logical continuity, for example, between describing a situation and its usual result, it's not an introductory phrase and there is no comma. In this case, it would often be possible to combine a "So" sentence with the preceding sentence.

✔ **Yes:**   *Don't touch the sensor. Also, make sure that the sensor is positioned vertically above the specimen holder.*

✔ **Yes:**   *Don't touch the sensor. Also don't touch the specimen holder.*

✔ **Yes:**   *So, now that you know this rule, you'll make fewer mistakes.*

✔ **Yes:**   *A car is a lot heavier than a motorcycle. So it consumes more fuel.*

or:

A car is a lot heavier than a motorcycle, so it consumes more fuel.

Commas are used before and after "for example"

Use commas before and after *for example*.

✔ **Yes:** *For example, this sentence has a comma.*

✔ **Yes:** *This sentence has a comma, for example.*

✔ **Yes:** *This sentence, for example, has two commas.*

Commas MUST NOT separate the subject from the predicate or the verb from the object

The following examples are just simple sentences. They're *not* a main clause plus another main clause, and they're *not* a main clause plus a subordinate clause.

✔ **Yes:** *A man of his abilities would always be successful.*

✔ **Yes:** *We thought that the machine would run out of fuel.*

Commas are used only before non-defining relative clauses

Add a comma before a non-defining relative clause. Non-defining relative clauses often begin with *which*.

Don't add a comma before a defining relative clause. Defining relative clauses often begin with *that*.

> ❶ **Important:** Make sure that you understand the different uses of *that* and *which*. For details, see *that / which* [252].

✔ **Yes:** *The program now imports the data, which may take several minutes.*

✔ **Yes:** *The program now imports the data that you've selected.*

✔ **Yes:** *This is an optional field, where you can enter some notes.*

✔ **Yes:** *This is the place where you must enter your name.*

Note that the presence or absence of a comma can make a great difference in meaning.

✔ **Yes:** *Most users of the program, who have little time, don't read the manual.*

(This means that most users don't read the manual because they have little time.)

✔ **Yes:** *Most users of the program who have little time don't read the manual.*

(This means that usually users do read the manual, but those who have little time don't.)

Usually, there's no comma before "that"

Usually, there's no comma before that.

✔ **Yes:** *It's important that you read all safety instructions before using the product.*

✔ **Yes:** *This is the answer that you've been looking for.*

Sometimes a that-clause follows an abstract noun like *answer, belief, demand, fact, knowledge, news, proposal, statement,* and so on. If the that-clause isn't necessary to complete the meaning of the sentence, mark the that-clause by commas.

✔ **Yes:** *The fact that the device is easy to use makes it suitable for beginners.*

(The short version of this sentence is: "The fact makes it easy for beginners." This doesn't make sense.)

✔ **Yes:** *The initial presumption, that specimen A was heavier than specimen B, was wrong.*

(The short version of this sentence is: "The initial presumption was wrong." This makes sense. The that-clause is just an addition to this statement.)

There's NO comma before "if"

Normally, there's no comma before *if*.

However, there's a comma *within* a sentence that begins with *If*.

✔ **Yes:** *Your car looks dirty if you don't wash it.*

✔ **Yes:** *If you don't wash your car, it looks dirty.*

There may be a comma before *if* when there's a list of if-clauses, or after an introductory phrase.

✔ **Yes:** *If you want to save your settings, if you want to print reports, or if you want to send reports by email, you need to buy the Pro version.*

✔ **Yes:** *Therefore, if you still don't know how to use commas, ...*

There's always a comma before "too" at the end of a sentence

There's always a comma before "too" at the very end of a sentence.

✔ **Yes:**   *The temperature of the engine is too hot.*

✔ **Yes:**   *You can learn how to use commas, too.*

2.6.6 Dashes

Don't confuse a dash with a hyphen (see *Hyphens* 183). Dashes are longer than hyphens.

There are two types of dashes: the en dash and the em dash. The en dash is based on the size of an uppercase *N*, and the em dash is based on the size of an uppercase *M*. So the em dash is longer than the en dash.

- hyphen

– en dash

— em dash

The en dash is a linking device that implies *to* or *and*.

The em dash is a separating device. It signals a pause slightly longer than a comma, or it interrupts the sentence for a purposeful digression.

When to use an en dash

Use an en dash to indicate continuing or inclusive numbers, such as dates, times, or reference numbers.

Also use an en dash as a minus sign.

✔ **Yes:** *August–September*

✔ **Yes:** *pages 10–20*

✔ **Yes:** *–4 °C*

When to use an em dash

Use an em dash to enclose or set off a parenthetic expression when commas and parentheses aren't emphatic enough.

Tip:
Sometimes it's better to split a sentence instead of using em dashes. Two short sentences are easier to read than one long sentence. See also *Avoid parentheses and nested sentences* 92 .

✔ **Yes:** *Use an em dash—if you can't avoid it—when commas aren't emphatic enough.*

✔ **Top:** *When commas aren't emphatic enough, you can use an em dash. However, try to avoid it. Instead, split the sentence into two sentences.*

How to enter dashes

On standard computer keyboards, there are no keys for dashes.

Some authoring tools automatically insert dashes when you type particular strings, such as two or three consecutive hyphens.

Tip:
If your authoring tool doesn't support dashes, you can also use the Special Characters Script included in the indoition Hotkey Script Collection for Writers and Translators to enter en dashes and em dashes (for more information, visit http://www.indoition.com).

To enter en dashes manually:

- On a computer that runs Windows: Hold down Alt and type 0150 on the numeric keypad.
- On a computer that runs Mac OS: Press Alt+Hyphen.
- On a computer that runs Linux: Press Ctrl+Shift+U, then type 2013, then press Enter.

To enter em dashes manually:

- On a computer that runs Windows: Hold down Alt and type 0151 on the numeric keypad.
- On a computer that runs Mac OS: Press Alt+Shift+Hyphen.
- On a computer that runs Linux: Press Ctrl+Shift+U, then type 2014, then press Enter.

The HTML code for an en dash is `&ndash;`.

The HTML code for an em dash is `&mdash;`.

The Unicode code point for the en dash is U+2013.

The Unicode code point for the em dash is U+2014.

When to add space characters before and after dashes

Most style guides recommend that you don't leave spaces on either side of en dashes and em dashes.

✔ **Yes:** *Use an em dash—if you can't avoid it—when commas aren't emphatic enough.*

Some writers use an open-set en dash (space character + en dash + space character) or an open-set em dash (space character+ em dash + space character) instead of a closed-set em dash. Although this is rather uncommon, particularly in American English, it isn't strictly wrong. Open-set en dashes instead of closed-set em dashes are also the predominant style in German and French typography, for example.

No matter which style you choose, use it consistently.

✔ **Yes:** *Open-set em dashes — like in this sentence — aren't wrong, but they are less common than closed-set em dashes.*

✔ **Yes:** *Alternatively, you can use an open-set en dash – as in German or French – instead of a closed-set em dash.*

2.6.7   Ellipses

Use an ellipsis (three dots) to indicate an omission.

However, don't use an ellipsis to indicate an idea that readers must complete by themselves. It's your job to do the thinking for the readers, so write down the solution

✘ **No:**   *Choose one of the primary colors red, green, ...*

✔ **Yes:**   *Choose one of the primary colors red, green, or blue.*

✔ **Yes:**   If the full list entry is "Primary colors red, green, blue": *From the list, select the entry* **Primary colors** *....*

Ellipsis special character

If you want to produce a document of premium typographic quality, note that most fonts have a special ellipsis character, which isn't on your keyboard. Some word processors automatically enter this character when you type three periods in a row. The ellipsis character is slightly smaller than three normal periods.

Three normal periods: ...
Ellipsis character: ...

Tip:
If your authoring tool doesn't support the ellipsis character, you can use the Special Characters Script included in the indoition Hotkey Script Collection for Writers and Translators to enter the correct character automatically when you type three periods in a row (for more information, visit *http://www.indoition.com*).

To enter an ellipsis character manually:

- On a computer that runs Windows: Hold down Alt and type 0133 on the numeric keypad.

- On a computer that runs Mac OS: Press Alt+Period.

- On a computer that runs Linux: Press Ctrl+Shift+ U, then type 2026, then press Enter.

The HTML code for the ellipsis special character is `&hellip;`.

The Unicode code point for the ellipsis special character is U+2026.

If you're using a font that doesn't support the ellipsis character, use three periods with no space characters between them.

Where to add space characters and punctuation marks

When the ellipsis represents *a part of a word*, don't add any space characters before and after the ellipsis.

✔ **Yes:** *In words that begin with the prefix pre..., don't use a hyphen.*

When the ellipsis represents *one or more complete words or sentences*, style guides differ in their recommendations on where you should add spaces and punctuation marks. Regardless of which style you choose, use it consistently.

The easiest solution is to treat the ellipsis exactly as you would treat the text that the ellipsis replaces:

At the beginning of a sentence fragment, add a space character after the ellipsis but no punctuation.

✔ **Yes:** *... Here we go.*

In the middle of a sentence, put a space character both before and after the ellipsis unless the ellipsis is followed by some punctuation, such as a comma.

✔ **Yes:** *This sentence explains ... and then continues.*

✔ **Yes:** *Click the item* **Print** *..., and then click* **OK.**

At the end of a sentence, add a space, then the ellipsis, and then closing punctuation (without an intervening space character).

✔ **Yes:** *This sentence explains ....*

✔ **Yes:** *What does this style guide tell you about headings, subheadings, subsubheadings, ...?*

If the ellipsis replaces one or more complete sentences, add a space plus the ellipsis without any closing punctuation.

✔ **Yes:** *This sentence explains how to use an ellipsis. ...*

✔ **Yes:** *This sentence explains how to use an ellipsis. ... And the next sentence continues.*

How to handle line breaks

Never break a line before the ellipsis. If necessary, break the line before the last word that precedes the ellipsis.

If you don't use the ellipsis special character but use three normal periods, make sure that your authoring tool doesn't insert any automatic line break between the periods.

2.6.8 Exclamation points

> Don't use exclamations in user assistance.
>
> Readers want guidance, but they don't like to be given orders.

In procedures, use full stops.

✗ No: *3. Click OK!*

✔ Yes: **3. Click OK.**

If you need to emphasize a word or statement, use a special character style (such as bold or italic), but don't use an exclamation point.

✗ No: *When you exit the program, your data is not(!) stored automatically.*

✔ Yes: **When you exit the program, your data is not stored automatically.**

        alternatively:

        When you exit the program, your data is NOT stored automatically.

For warnings, it's OK to use an exclamation point as part of a warning symbol. However, don't rely on an exclamation point alone. Its visibility isn't strong enough. Add an appropriate signal word, such as *Caution*, *Warning*, or *Danger* (see *Writing warnings* 72).

✗ No: *Risk of fire! Don't smoke while filling up the tank!*

✔ Yes: ⚠ **Caution: Risk of fire. Don't smoke while filling up the tank.**

Related rules

Writing warnings 72

2.6.9 Hyphens

Don't confuse a hyphen with a dash (see *Dashes* [177]).

To enter a hyphen, use the standard hyphen on your keyboard.

Nouns

Don't hyphenate compound nouns.

Usually, write compound nouns as two separate words.

Only write compound nouns as one word if this spelling exists in a dictionary. See also *FAQ: Standard terms and phrases* [261].

✘ No: *web-page, web-site, mouse-click*

✘ No: *webpage, website, mouseclick*

✔ Yes: *web page, web site, mouse click*

✔ Yes: *keyboard, database*

Adjectives

Always use hyphens in compound adjectives. Hyphens are optional here but help the reader to analyze the sentence more quickly. Also, they can sometimes eliminate ambiguities.

✔ Yes: *top-quality cable*

✔ Yes: *low-resolution monitor*

✔ Yes: *high-level-language compiler*

✔ Yes: *decision-making process*
(but: the process of decision making)

✔ Yes: *well-designed user interface*
(but: the user interface was well designed)

✔ Yes: *easy-to-use program*
(but: the program is easy to use)

✔ Yes: *little used car*
(means: a small second-hand car)

little-used car
(means: a car that wasn't used very much)

✔ **Yes:** *a heavy-metal detector*
(means: a device that detects heavy metals)

a heavy metal detector
(means: a heavy detector that detects all sorts of metals—not just heavy metals)

In compound adjectives in which one of the elements is an open compound, use an en dash instead of a hyphen (see *Dashes* 177). Don't replace the space character with a second hyphen.

✔ **Yes:** *Windows 7–compatible software*

✔ **Yes:** *dialog box–like window*

You can also use hyphens with successive compound modifiers. However, it's often better to rephrase such sentences.

✔ **Yes:** *pre- and postpaid*

✔ **Top:** *prepaid and postpaid*

✔ **Yes:** *There are 10-, 25-, and 50-meter cables available.*

✔ **Top:** *There are 10-meter, 25-meter, and 50-meter cables available.*

even better:

Cables are available in lengths of 10 meters, 25 meters, and 50 meters.

Don't hyphenate generic descriptors of products because they can also be seen as part of a compound noun.

✔ **Yes:** *data processing unit*

Don't hyphenate phrases with an adverb in front of an adjective. The ending "...ly" already makes it clear that the adverb relates to the following adjective, so a hyphen would be redundant.

However, do use a hyphen if the adverb is part of a longer compound.

✔ **Yes:** *highly complex software*

✔ **Yes:** *a remotely controlled machine*

✔ **Yes:** *formally-agreed-upon format*

Words with prefixes

In general, don't hyphenate words with prefixes. There are only a few cases where a hyphen is needed.

Don't hyphenate

- *anti...*
  Examples: *antialiasing, antiseptic*

- *auto...*
  Examples: *autocompletion, autocatalytic*

- *co...*
  Examples: *coexistence, coprocessor*

- *de...*
  Examples: *deactivate, decoder*

- *multi...*
  Examples: *multicolor, multidimensional*

- *non...*
  Examples: *nonexistent, nontoxic*

- *pre...*
  Examples: *preface, prefilter*

- *post...*
  Examples: *postprocessing, postproduction*

- *re...*
  Examples: *restart, reaccept*

- *sub...*
  Examples: *submenu, subtotal*

- *un...*
  Examples: *undelete, uncluttered*

Hyphenate

- if the prefix is *self* or *all*
  Examples: *self-controlled, self-extracting, self-running, all-inclusive*

- if the hyphen helps to pronounce the word correctly
  Examples: *non-native, re-evaluate, multi-instrument, ultra-adaptive*

- if the hyphen is needed to avoid misinterpretation
  Examples: *reform* as opposed to *re-form*, *remark* as opposed to *re-mark*, *recover* as opposed to *re-cover*

- if the compound uses a double prefix
  Example: *sub-submenu*

- if the compound contains another compound
  Example: *non-security-related* (better: *unrelated to security*; *not related to security*)

- if the base is a number
  Examples: *pre-2000*, *post-1945*
- if the base is capitalized
  Examples: *pro-American*, *non-Asian*, *mid-September*
- if you combine an initial letter with a word
  Examples: *I-beam*, *T-shirt*, *x-axis*, *e-learning*

Note:
According to this rule, it's OK to write *e-mail*. However, it's becoming increasingly more common to write *email* without a hyphen.

Words with suffixes

There are no universal, reliable rules here. When in doubt, consult a dictionary.

If you don't find a word in a dictionary, use the alternative that's easier to read.

See also *FAQ: Standard terms and phrases* 261

Numbers

Use hyphens to combine numeral-unit adjectives.

Don't use hyphens with percentages and symbols.

Don't use hyphens with page numbers, chapter numbers, figure numbers, table numbers, example numbers, version numbers, slot numbers, and so on.

Use hyphens to combine fractions.

Use hyphens to write compound numbers from 21 to 99. (Note: It's generally better to write these numbers as numerals. See *Numbers* 188.)

✔ **Yes:**   *10-meter cable*
✔ **Yes:**   *12-V power supply*

✔ **Yes:**   *10 m*
✔ **Yes:**   *12 V*
✔ **Yes:**   *50 percent increase*
✔ **Yes:**   *3% decline*
✔ **Yes:**   *45° angle*
✔ **Yes:**   *€50 bill*

✔ **Yes:**   *page 25*
✔ **Yes:**   *figure 7*

✔ **Yes:** *table 12*
✔ **Yes:** *example 5*
✔ **Yes:** *slot 3*

✔ **Yes:** *two-thirds*
✔ **Top:** *2/3*

✔ **Yes:** *sixty-eight*
✔ **Top:** *68*

Related rules
Dashes 177

2.6.10 Numbers

> **1** In general, use numerals. They're shorter and easier to grasp than numbers spelled out as text.
>
> **2** Be consistent and parallel in the way that you handle closely related sentences and paragraphs. When comparing values, or when referring to the same base units, write these numbers the same way.
>
> When referring to different base units, use a different notation. Spell out the smaller numbers and write the larger numbers as numerals.
>
> **3** Don't use a numeral at the beginning of a sentence. Try to rephrase the sentence.

1

✘ No: *There are 50 apples but only three pears.*

✔ Yes: *There are fifty apples but only three pears.*

✔ Top: *There are 50 apples but only 3 pears.*

2

✔ Yes: *You'll need five cables: three cables that are 5 m long, and two cables that are 10 m long.*

(The numbers for the numbers of items are both spelled out: "five cables" and "three cables." The lengths are both given as numerals: "5 m long" and "10 m long.")

✔ Yes: *two 64-GB USB drives*

(The number of drives is spelled out: "two drives." The storage capacity is given as a numeral because it refers to a different base unit: "64 GB.")

3

✘ No: *1024 is a good value to start with.*

✔ Yes: *A good value to start with is 1024.*

Millions, billions, and trillions

Billion, trillion, and some larger numbers have different meanings in different countries. For example, a billion is 10^9 in the United States but 10^{12} in many other countries.

For this reason, use numerals or powers of 10 when talking about numbers greater than one million.

Percentages

In text, use the word *percent*.

In tables or other places where space is tight, use the percent sign (%).

Don't put a space character before the percent sign.

✔ **Yes:** *There was a 50 percent increase of ....*

✔ **Yes:** *50%*

Units of measure

Put a space character between numbers and units of measure.

Don't put a period after the unit of measure unless the unit of measure stands at the end of a sentence.

✔ **Yes:** *8 GB*

✔ **Yes:** *12 V*

✔ **Yes:** *10 m*

✔ **Yes:** *1-1/5 inches*

 Related rules

Hyphens 183

2.6.11 Quotation marks

1 Use quotation marks to indicate that you're quoting a person, or from a publication, or from a product's user interface.

2 When possible, use a special character style rather than quotation marks to specify terms and quotations that are from a user interface.

Never use both a special character style *and* quotations marks at the same time.

3 Don't use quotation marks around a term because you're afraid that some users might not understand this term. Instead, use a term that all users understand, or describe what you mean in more detail.

In informal texts, it can be acceptable to add quotation marks around terms if these terms aren't used literally. However, it's often better to rephrase the sentence.

4 Don't use quotation marks around terms that follow *so-called*. Often, the best solution is to leave out the term *so-called* altogether.

1

✔ **Yes:** *If the message "Replace existing file?" appears, click OK.*

2

✘ **No:** *Click the "Print" button.*

    (uses both, quotation marks *and* a special, bold character style)

✔ **Yes:** *Click the "Print" button.*

    (only uses quotation marks)

✔ **Top:** *Click the Print button.*

    (only uses a special, bold character style)

3

✘ **No:** *You can also "wrzlprmf" the cable.*

✔ **Yes:** *You can also connect the cable so that it ....*

4

✘ **No:** The so-called "AB connector" connects device A with device B.

✔ **Yes:** The so-called AB connector connects device A with device B.

✔ **Top:** The AB connector connects device A with device B.

Quotation marks and punctuation

In the United States and in Canada, place periods and commas inside the quotation marks. In Europe, place periods and commas outside the quotation marks.

✔ **Yes:** This is "fool-proof design." (US, Canada)

✔ **Yes:** This is "fool-proof design". (Europe)

✔ **Yes:** As this is "fool-proof design," there should be no problem. (US, Canada)

✔ **Yes:** As this is "fool-proof design", there should be no problem. (Europe)

Question marks, exclamation points, dashes, colons, and semicolons go either inside or outside the quotation marks, depending on whether they belong to the quoted material or to the sentence as a whole.

✔ **Yes:** The following characteristics are essential for "fool-proof design": ...

✔ **Yes:** The following message appears: "Risk of electric shock!"

Never add double punctuation marks, such as a question mark or exclamation point followed by a period, or a period followed by another period.

✘ **No:** The following message appears: "Risk of electric shock!".

✔ **Yes:** The following message appears: "Risk of electric shock!"

Typographical conventions

For normal quotations, use double quotation marks, not single quotation marks. If a quotation includes another quotation, use single quotation marks for the inner quotation.

With the invention of the typewriter, a neutral, straight form of quotation marks was created so writers could use the same typewriter key for both opening and closing quotation marks. These straight quotation marks are available on your computer keyboard.

In contrast to straight quotation marks, typographic quotation marks look like the numbers 66 and 99. This is why they're also called curly quotation marks.

> **ⓘ Important:** Other languages than English use different conventions.

To give your document a truly professional appearance, use curly quotation marks rather than straight quotation marks (exception: don't replace straight quotation marks when describing or citing program source code).

In English, quotation marks look like the following:

straight double quotation marks:	**"text"**
curly double quotation marks:	**"text"**
straight single quotation marks:	**'text'**
curly single quotation marks:	**'text'**

How to enter curly quotation marks

On standard computer keyboards, there's a key for straight quotation marks, but there is no key for curly quotation marks.

Today, many authoring tools automatically insert curly quotation marks when you type straight quotation marks.

Tip:
If your authoring tool doesn't support curly quotation marks, you can also use the Special Characters Script included in the indoition Hotkey Script Collection for Writers and Translators to enter them very conveniently (for more information, visit *http://www.indoition.com*).

To enter curly quotation marks manually, proceed as follows.

On a computer that runs **Windows**:

- Opening double quotation mark: Hold down Alt and type 0147 on the numeric keypad.
- Closing double quotation mark: Hold down Alt and type 0148 on the numeric keypad.
- Opening single quotation mark: Hold down Alt and type 0145 on the numeric keypad.
- Closing single quotation mark: Hold down Alt and type 0146 on the numeric keypad.

On a computer that runs **Mac OS**:

- Opening double quotation mark: Press Option+[.
- Closing double quotation mark: Press Option+Shift+[.
- Opening single quotation mark: Press Option+].
- Closing single quotation mark: Press Option+Shift+].

On a computer that runs **Linux**:

- Opening double quotation mark: Press Ctrl+Shift+U, then type 201c, then press Enter.
- Closing double quotation mark: Press Ctrl+Shift+U, then type 201d, then press Enter.
- Opening single quotation mark: Press Ctrl+Shift+U, then type 2018, then press Enter.
- Closing single quotation mark: Press Ctrl+Shift+U, then type 2019, then press Enter.

In **HTML**, use the following codes:

- Opening double quotation mark: “
- Closing double quotation mark: ”
- Opening single quotation mark: ‘
- Closing single quotation mark: ’

The **Unicode** code points are:

- Opening double quotation mark: U+201c

- Closing double quotation mark: U+201d
- Opening single quotation mark: U+2018
- Closing single quotation mark: U+2019

2.6.12   Semicolons

1 Use semicolons to link closely related clauses or to join contrasting statements.

Be sure that the statement to the right of the semicolon is a complete sentence. If it's not, use a comma or colon instead of a semicolon, or rephrase the sentence.

If it isn't important to have both statements within the same sentence, it's often better to split the sentence into two sentences (see also *Make short sentences* 88).

2 You can also use a semicolon to visualize the structure of a sentence:

- before words such as *then*, *however*, *yet*, *consequently*, and *furthermore*
- in a series if one or more of the clauses contain commas

1

✔ **Yes:**   *Adding a new device driver is easy; removing a device driver can be difficult.*

✘ **No:**   *Add one liter of water if you're painting on wood; two liters of water if you're painting on concrete.*

✔ **Yes:**   *If you're painting on wood: Add one liter of water.*
If you're painting on concrete: Add two liters of water.

✔ **Yes:**   *Setup is easy; the full procedure only takes about an hour.*

✔ **Top:**   *Setup is easy. The full setup procedure only takes about an hour.*

2

✘ **No:**   *You can print photos on standard paper, however, we recommend using premium photo paper.*

✔ **Yes:**   *You can print photos on standard paper; however, we recommend using premium photo paper.*

✘ **No:**   *Possible entries are: the colors red, green, or blue, the line widths thin, medium, or thick, and the line styles solid or dashed.*

✔ **Yes:**   *Possible entries are: the colors red, green, or blue; the line widths thin, medium, or thick; and the line styles solid or dashed.*

✔ **Yes:** *Possible entries are: the color (red, green, blue), the line width (thin, medium, thick), and the line style (solid, dashed).*

✔ **Top:** *Possible entries are:*

- *colors (red, green, or blue)*

- *the line width (thin, medium, or thick)*

- *the line style (solid or dashed)*

2.7 FAQ: Grammar and word choice

Even small errors affect your credibility and undermine the quality of your product.

Don't underestimate the importance of correct grammar and word choice. Flawless instructions are much more trustworthy than those with many errors.

In everyday speech, many words are used interchangeably. However, when you write instructions and technical documentation, choosing one word as opposed to another can make a vital difference.

> **Important:** Use electronic grammar checkers with care. They can identify many mistakes, but they can't find them all. Electronic spelling checkers, grammar checkers, and other language tools don't work as a substitute for editing and proofreading by a human.

FAQ

When writing a text, you probably don't want to waste your time browsing bulky grammar reference manuals and textbooks. For this reason, we've compiled quick answers to the most frequent questions that arise when writing instructions and technical documents.

Tip:
The listed topics include not only frequently asked questions but also frequently made mistakes. Even if you don't have any particular question now, take the time to skim the topics in this section for details that you might not be aware of.

The terms are sorted alphabetically. If one topic covers several terms, the more common term is listed before the more uncommon term. If you're looking for a specific word and don't find it, look for it in the index.

Related rules

FAQ: Standard terms and phrases 261

2.7.1 accurate / precise

Use *accurate* when you mean *correct, free from error*, or *true*.

Use *precise* when you mean *minutely exact down to a couple of decimal points* or *sharply defined*. *Precise* also refers to conformance to a strict standard or pattern. Think of *precise* as a very narrow range of tolerance.

✔ **Yes:** *This instrument is very accurate. It always shows the correct value.*

✔ **Yes:** *This instrument is precise. It shows the value with a precision of 10 decimals.*

✔ **Yes:** *You can't be too accurate, but you can be too precise if you exceed the number of significant figures in your calculations.*

2.7.2   allow / enable

People *allow* things, but things *enable* people.

Use *allow* when you mean *permit* or *consent to*. Also use *allow* when you mean *to allocate a certain amount*.

Use *enable* when you mean *provide the means or power*, or *make something possible or easy*. In this case, it's often better to use *can* or *let*.

✔ **Yes:**   You're allowed to make one backup copy of the program.

✔ **Yes:**   The trial version of the program allows you to create only documents that have 3 pages or less.

✔ **Yes:**   Leaving some space in the drum allows for expansion of the liquid when the liquid is hot.

✘ **No:**   This program allows you to hack your competitor's web site.

✔ **Yes:**   This program enables you to hack your competitor's web site.

✔ **Top:**   With this program, you can hack your competitor's web site.

2.7.3 Alphabetical order

> When sorting entries alphabetically, sort them word by word, not letter by letter.
>
> So don't ignore space characters.

✖ No: *Enter command*
            entering text
            Enter key

✔ Yes: **Enter command**
            Enter key
            entering text

2.7.4 among / between

Use *among*
- when referring to more than two things
- when the number is unspecified

Use *between*
- when referring to exactly two things
- when referring to a discrete number of things

✔ **Yes:** *There are some green keys among the keys on the keyboard.*

✔ **Yes:** *You can choose between steak, chicken, and fish.*

✔ **Yes:** *You can switch between open programs.*

(Note: You always switch between two programs.)

✔ **Yes:** *The yellow indicator light is located between the green indicator light and the red indicator light.*

✔ **Yes:** *Hold the needle between your index finger and thumb.*

✔ **Yes:** *Choose any number between 1 and 10.*

(Note: You have 10 options to choose from, so the number of options isn't unspecified.)

2.7.5   amount / number

Use *amount* when referring to abstract, inseparable, continuous, or otherwise uncountable quantities.

Use *number* when referring to countable objects.

Tip:
Instead of the phrase *a large amount of*, use the shorter word *many*. Instead of the phrase *a large number of*, use the shorter phrases *much, a lot of*, or *lots of* (see also *many / much / a lot of / lots of* 244).

✔ **Yes:**   *The amount of information depends on ....*
✔ **Yes:**   *The number of options depends on ....*

✔ **Yes:**   *a large amount of information*
✔ **Top:**   *much information* (formal)
            a lot of information (informal)
            lots of information (informal)

✔ **Yes:**   *a large number of users*
✔ **Top:**   *many users* (formal)
            a lot of users (informal)
            lots of users (informal)

Related rules

many / much / a lot of / lots of 244

2.7.6   and / as well as / plus

Usually, use *and*.

However, *as well as* and *plus* can make it easier for readers to predict and analyze the structure of the rest of a sentence.

Use *as well as* to set off different items in a list.

Avoid *as well as* when it might be confused with *as good as*, especially if you're writing for an international audience.

✘ No: You can use the machine to cut wood as well as steel.

✘ No: You can use the machine to cut wood plus steel.

✔ Yes: You can use the machine to cut wood and steel.

✔ Yes: You can use the machine to cut wood, steel, and plastics.

✔ Yes: We sell computers, software, and T-shirts.

✔ Yes: We sell computers and software, as well as T-shirts.

(Note: There are two different classes of goods: technical products and clothing.)

✔ Yes: The form lets you enter your name and other personal data.

(Note: After the word *and*, this sentence might also continue very differently. Example: "... and if you aren't careful you might end up on a spammer's list.")

✔ Top: The form lets you enter your name, as well as other personal data.

(Note: Here, after *as well as*, the reader can expect a noun. This makes the sentence more predictable and easier to read.)

✔ Yes: The kit includes two cables and a selection of spare parts.

✔ Top: The kit includes two cables plus a selection of spare parts.

(Note: The word *plus* clearly signals that you're talking about a list of things. If you use *and*, the sentence might also continue very differently. Example: "... and is one of the best on the market.")

✔ Yes: You can write manuals as well as online help files.

(Note: This sentence is ambiguous. Does it mean that you can write both manuals *and* help files? Or does it mean that you're good at writing manuals and equally good at writing online help files?)

✔ Top: You can write manuals plus online help files.

2.7.7 assume / presume

Assume and presume are often used interchangeably. However, there's a slight difference:

- To *assume* means to take for granted as the basis of argument or action. There doesn't have to be any proof for the assumption; it can be arbitrary. You can replace the word *assume* with *suppose*.

- To *presume* means to assert that something is true without yet having complete evidence (hence the prefix "pre..."). However, there's a belief that at least some evidence exists.

So there's a stronger element of postulation or hypothesis in *assume* than in *presume*.

The nouns are *presumption* and *assumption*.

✔ **Yes:**   *Let's assume that two users log in who both have the same name.*

✔ **Yes:**   *To start the iterative process, let's presume that a typical value for the average room temperature is 20° Celsius.*

2.7.8   because / since / as

> Use *because* to refer to a reason.
>
> Use *since* only to refer to the passage of time. This avoids ambiguities and prevents misinterpretation.
>
> Don't use *as* as a synonym for *because*. The word *as* can be ambiguous and has so many meanings that it's problematic, especially for an international audience.

✘ No: Since the paper is missing, you can't print.

✘ No: As the paper is missing, you can't print.

✔ Yes: Because the paper is missing, you can't print.

✔ Yes: Since Monday, the paper has been missing.

✔ Yes: Because we installed the program, we've saved a lot of time.

　　　　(means: "We've saved a lot of time. The reason for this is the installed program.")

✔ Yes: Since we installed the program, we've saved a lot of time.

　　　　(means: "We've saved a lot of time. This happened after we installed the program.")

Related rules

while / as / whereas / although 258

Feel free to start sentences simply 94

2.7.9 big / great / large / little / small

> The words *big* and *little* have somewhat emotional connotations.
>
> In technical documentation, use *large* and *small* instead.
>
> Use **great** only for things that are especially large. Don't use *great* as a synonym for *very good* or *excellent* (see also *Don't make judgments* 41).

✘ No: *The device has a big monitor.*
✔ Yes: **The device has a large monitor.**

✘ No: *Press the little red button.*
✔ Yes: **Press the small red button.**

✘ No: *A great way of doing this ....*
✔ Yes: **A good way of doing this ....**

✘ No: *This is a great example.*
✔ Yes: **This is an excellent example.**

2.7.10   can / may / might / must / should

Always be as precise as possible. The incorrect or vague use of the terms *can*, *may*, *might*, *must*, *should*, *could*, and so on is one of the most frequent causes for misinterpretation.

1 Use *can* when you mean the ability or power to do something.

2 *May* and *might* both indicate possibility or probability. *Might* suggests a somewhat lower probability than *may*.

Don't use *may* to imply the ability to do something. In this case, use *can*.

Don't use *may* to imply the permission to do something. In this case, use *allowed to*.

Tip:
Phrases with *you* include *can* more often than *may* ("You can ....").

3 Use *must* to describe a user action that's required. If you feel that the word *must* is too strong or impolite because it implies an obligation, rephrase your sentence as an instruction or use *need to* or *have to*.

Note:
In American English, don't use *shouldn't* instead of *mustn't* because you think that *mustn't* sounds too British. Use *must not* instead, which is unambiguous for an international audience.

4 Use *should* only to describe a user action that's recommended but optional. However, try to avoid the word *should* altogether because it always conveys an element of doubt. Instead, clearly tell the reader what to do, or clearly mark your sentence as a recommendation.

Never use the word *shall* in technical documentation.

1

✘ **No:**   *You may use a spoon or a fork.*

✔ **Yes:**   *You can use a spoon or a fork.*

✘ **No:**   *You may use the program to write a manual.*

✔ **Yes:**   *You can use the program to write a manual.*

2

✔ **Yes:**   *You may be right.*

✔ **Yes:**   *Print quality may be poor if you use cheap paper.*

✔ **Yes:**   *The camera is shock proof, but it might break if you drop it from a height of more than 1.5 meters.*

✘ **No:**   *The remote control may also use a frequency of 140 MHz.*

(This is Ambiguous: It could mean that (a) you're allowed to use a frequency of 140 MHz; (b) the device is able to send at 140 MHz if you set it up to do so; (c) it might happen that the device sends at 140 MHz at its own discretion.)

✔ **Yes:**   *It's also allowed to use a frequency of 140 MHz for the remote control.*
(If you mean that law permits you to use this frequency.)

The remote control can also use a frequency of 140 MHz.
(If you mean that the device is able to send at this frequency.)

The remote control might also use a frequency of 140 MHz.
(If you mean that the device chooses the frequency automatically.)

✘ **No:**   *You may print the report, or you may save it to a file.*

✔ **Yes:**   *You can print the report, or you can save it to a file.*

3

✔ **Yes:**   *The fluid must pass the valve.*

✔ **Yes:**   *If the function doesn't work, you must contact support.*

✔ **Top:**   *If the function doesn't work, you need to contact support.*

✔ **Top:**   *If the function doesn't work, contact support.*

4

✘ **No:**   *You should make a backup copy of the file.*

✔ **Yes:**   *We recommend that you make a backup copy of the file.*

✔ **Top:**   *Make a backup copy of the file.*

✖ **No:**   You should get a reply within 24 hours.

✔ **Yes:**   You will usually get a reply within 24 hours.

✔ **Top:**   We usually reply within 24 hours.

---

must / must not

For many writers who speak English as a second language, the use of *must* and *must not* is a dangerous pitfall. In some languages, the literal translation of *must not* means *need not*. In English, *must not* and its contraction *mustn't* mean that there's an **obligation not to do something**.

✔ **Yes:**   You must not smoke while filling up the tank.

✔ **Yes:**   You must not use a 110-volt device on a 220-volt outlet.

---

cannot / can not

Use *cannot* or its contraction *can't* if you mean *is not able to*. Here, the word *not* negates the word *can*.

Use *can not* if the word *not* relates to the action following the word *can*, not to the word *can*.

✔ **Yes:**   You cannot / can't start the compressor while the engine is running.

(means: It isn't possible to start the compressor while the engine is running. It won't work.)

✔ **Yes:**   You can not start the compressor while the engine is running.

(means: You don't have to start the compressor while the engine is running. It's not required, although it may be more common or recommended.)

---

can / could

The word *could* conveys an element of doubt, which is something that you must avoid in user assistance (see *Be specific* 23). For this reason, don't use *could* when you mean *can*, *does*, or *will*.

Only use *could* as the past tense of *can*. In most cases, however, it's better to use the present tense (see *Use the present tense* 38).

✖ **No:**   If you follow this rule, this could improve your documents.

✔ **Yes:**   If you follow this rule, this improves your documents.

✘ No: *Instead of using the menu, you could also use a keyboard shortcut to format the text.*

✔ Yes: *Instead of using the menu, you can also use a keyboard shortcut to format the text.*

✔ Yes: *The program couldn't find your name in its database.*

✔ Top: *The program can't find your name in its database.*

Related rules

Be specific 23

2.7.11   check / control

> Use *check* when you mean *to examine* or *to make sure that something is correct, safe, or suitable*.
>
> Use *control* when you mean *to direct, to operate,* or *to limit something*.

✔ **Yes:** *You can check the brightness of the display by pressing the Info key.*

(means: You can read a number there, but you can't change the brightness.)

✔ **Yes:** *You can control the brightness of the display by pressing the Settings key.*

(means: You can change the brightness there.)

✘ **No:** *Control whether all wires are connected correctly.*

✔ **Yes:** *Check whether all wires are connected correctly.*

✔ **Yes:** *The slider allows you to control the voltage.*

2.7.12   compare to / compare with

Use *compare to* to point out either the similarities OR the differences between two things. *Compare to* likens two things. It puts them in the same category.

Use *compare with* to point out the similarities AND the differences. In *comparing with* something, one finds or discusses both things that are alike and things that are different.

✔ **Yes:**   *Scientists sometimes compare the human brain to a computer.*

✔ **Yes:**   *Compared to aluminum, steel is harder.*

✔ **Yes:**   *When comparing version 2 with version 1, you'll notice ….*

✔ **Yes:**   *Compare the second picture with the original. Do you see the difference?*

2.7.13   comprise / compose / constitute

A whole *comprises* parts.

Compose and *constitute* refer to the parts that make up the whole.

Don't use *is comprised of*.

Tip:
Even some native speakers of English have trouble with the verb *to comprise*. If possible, use trouble-free synonyms such as *consist of, include*, or *contain*.

✔ **Yes:**   *The office suite comprises a word processor, a spreadsheet application, and a database.*

✔ **Top:**   *The office suite includes a word processor, a spreadsheet application, and a database.*

✔ **Yes:**   *The manual comprises seven chapters.*

✔ **Top:**   *The manual consists of seven chapters.*

✔ **Yes:**   *The CPU, the hard disk, the keyboard, and the display constitute the device.*

✔ **Top:**   *The device consists of:*

- *CPU*
- *hard disk*
- *keyboard*
- *display*

✔ **Yes:**   *The training is composed of 6 lessons.*

✔ **Top:**   *The training includes 6 lessons.*

2.7.14   content / contents

> Use *content* when you mean *substance*, *information*, or *valuable data*.
>
> Use *contents* when you mean individual items inside something that can be picked out and listed. You can also use *contents* when you mean things that are written in a book, magazine, letter, or other document.

✔ **Yes:**   *The database has no content.*

  (means: There may be many bits and bytes stored in the database, but they don't make much sense.)

✔ **Yes:**   *The database has no contents.*

  (means: The database is actually empty and doesn't contain any data.)

✔ **Yes:**   *The sugar content is 50%.*

✔ **Yes:**   *Writers know that producing good content takes a lot of work.*

✔ **Yes:**   *The contents of the document are confidential.*

✔ **Yes:**   *The chapters are listed in the table of contents.*

2.7.15   continual / continuous

> Use *continual* if you want to say that something happens repeatedly.
>
> Use *continuous* if you want to say that something happens in one uninterrupted sequence.

✔ **Yes:**   *There's a continual sound.*

(means: Repeatedly, there's a sound. It stops and then starts again.)

✔ **Yes:**   *There's a continuous sound.*

(means: There's a sound that doesn't stop.)

✔ **Yes:**   *If you continually get error messages, please contact support.*

✔ **Yes:**   *There are continual problems with the engine.*

✔ **Yes:**   *We provide continuous support, 24 hours, 7 days a week.*

✔ **Yes:**   *The device transmits a continuous high-frequency signal.*

✔ **Yes:**   *A continuous row of items.*

2.7.16   cost / price / value / worth

Cost is what it takes in terms of money or resources to produce a particular product or service.

Price is the amount asked by the seller or paid by the buyer. If the price is higher than the cost, the seller makes a profit.

Value is the relation of the price to a recognized standard.

Worth is the relation of the price to a buyer's need or desire.

✔ **Yes:**   Production costs have remained stable.

✔ **Yes:**   We sell the product for a price of 50 dollars.

✔ **Yes:**   The product's many features give you a great value for your money.

✔ **Yes:**   The worth of the product comes from the increase in efficiency and the improvement of quality.

2.7.17   data / information

Don't use the terms *data* and *information* interchangeably.

- *Data* is a collection of raw material and facts. Data in itself is useless.
- When data is processed and structured in a given context, it becomes useful *information*.

Note:
Data can be used both in singular or plural form ("data is ...", "data are ..."). However, use the singular form, which is becoming increasingly more common.
Information is always used in singular form ("this information is ..."). There are no such forms as "datas" or "informations."

✔ **Yes:**   The program collects data from various sources.

✔ **Yes:**   You can then process this data to gain information on how ....

2.7.18 definite / definitive

> Use *definite* when you mean *precise, clearly decided and specific*, or *having distinct or certain limits*.
>
> Use *definitive* when you mean *conclusive* or *authoritative and exhaustive*.

✔ **Yes:** *There's been a definite increase in the number of support calls.*

✔ **Yes:** *There's a definite link between temperature and pressure.*

✔ **Yes:** *Arrange a definite date for the interview.*

✔ **Yes:** *The ISO standard is the definitive guideline for your work.*

✔ **Yes:** *This manual is your definitive guide to using DemoSoft.*

2.7.19 different from / different than

In comparisons, use *different from*.

Avoid *different than* because sentences that use *different than* are often hard to read.

However, it's OK to use *differently than*.

✖ **No:**   The result of the first operation is different than the result of the second operation.

✔ **Yes:**   The result of the first operation is different from the result of the second operation.

✔ **Yes:**   If the result is different from what you've expected, verify your settings.

✔ **Yes:**   The handling of this instrument is different from the handling of similar instruments.

✔ **Yes:**   Use instrument B differently than instrument A.

2.7.20   discreet / discrete

Use *discreet* when you mean *unobtrusive, unnoticeable,* or *careful not to do or say anything that's confidential or that could upset someone.*

Use *discrete* when you mean *separate, distinct, consisting of distinct or unconnected elements,* or *taking on or having a finite or countably infinite number of values.*

Tip:
The word *discrete* is more likely to appear in technical content.

✔ **Yes:**   You can find the serial number on a discreet label on the back of the device.

✔ **Yes:**   The articles are shipped in a discreet package.

(means: There's no label on the package that indicates what's in the package.)

✔ **Yes:**   The articles are shipped in discrete packages.

(means: Each article is shipped in a separate package.)

✔ **Yes:**   Enter one discrete number into each line.

✔ **Yes:**   The production line consists of several discrete sections, which are all controlled independently.

✔ **Yes:**   The function inserts a discrete random variable.

2.7.21 distinct / distinctive

Use *distinct* when you mean *clear, dedicated,* or *distinguishable.*

Use *distinctive* when you mean *of a nature that helps to distinguish a person or thing.*

✔ **Yes:** The product will give you a distinct competitive advantage.

✔ **Yes:** The design is very distinctive.

2.7.22   do / make / perform

Use *do* when you want to imply continuity.

Use *make* when you want to imply creation.

Use *perform* when you want to imply action.

Tip:
Often, the best solution is to use another, more direct verb altogether.

✔ **Yes:**   *They do an excellent job.*

✔ **Yes:**   *Make your settings.*

✔ **Yes:**   *Perform server administration outside local business hours.*

✔ **Top:**   *Administrate the server outside local business hours.*

2.7.23 each / every

1 Sometimes, *each* and *every* have the same meaning, but often they aren't exactly the same.

- Use *each* to emphasize individuality. *Each* expresses the idea of "one by one."

  Each can be followed by *of*. However, when possible omit the *of* to make your text more concise.

- *Every* refers to things or people in a group or in general. You can often replace *every* with *all*.

  Every is also always used to say how often something happens.

- When talking about two things, you must always use *each* or *each of*.

2 When used as a subject, *each* and *every* are always singular. The use of *each* or *every* clearly indicates you're considering individuals.

1

✔ **Yes:** Each of the computers that we produce has been tested.
Each computer that we produce has been tested.
Each has been tested.

✔ **Yes:** Every computer needs a power supply.

✔ **Yes:** Check the cable each time before you switch on the hairdryer.

✔ **Yes:** Hold one bar with each hand.

✔ **Yes:** We update the web site every day.

2

✘ **No:** Each of the computers have been tested.
✔ **Yes:** Each of the computers has been tested.

✘ **No:** Every computer need a power supply.
✔ **Yes:** Every computer needs a power supply.

2.7.24   effective / efficient

Use *effective* when you mean *producing a desired effect.*

Use *efficient* when you mean *productive* or *with little waste.*

✔ **Yes:**   *Cooling a desktop computer with liquid helium is effective, but it isn't efficient.*

(Helium cools very well, so it's effective. However, helium is more expensive than other ways of cooling a computer, so helium isn't efficient.)

✔ **Yes:**   *An efficient power supply has a higher effective power output than an inefficient power supply.*

2.7.25 fewer / less

Use *fewer* when referring to countable objects.

Use *less* when referring to abstract, inseparable, continuous, or otherwise uncountable quantities.

✔ **Yes:**   *You need fewer components to assemble device A than you need to assemble device B.*

✔ **Yes:**   *You need less time to assemble device A than you need to assemble device B.*

✔ **Yes:**   *The fewer devices you connect, the less cable you need.*

(Applies if the cable is on a drum and you need to cut it off.)

✔ **Yes:**   *The fewer devices you connect, the fewer cables you need.*

(Applies if you already have discrete, ready-to-use, short cables.)

2.7.26   figure / graphic / image / picture

If you use figure titles, use the general term *figure* both in the figure titles and when referring to figures.

If you don't use figure titles, when talking about a figure in your document, use the term that describes the type of figure as closely as possible. Use:

- *picture* (don't use the specific terms *photograph*, *screenshot*, *screen capture*, and *screen dump*)
- *drawing*
- *flow chart*
- *diagram*
- *illustration*

Use *image* only as a general term if you can't be more specific. Don't use *graphic*.

✔ **Yes:**   *Figure 4.2 shows ….*

✔ **Yes:**   *In the picture, you can see ….*

2.7.27   further / farther

Use *further* to refer to extent, degree, or time.

Use *farther* only to refer to physical distance.

✔ **Yes:**   *For further information, see ....*

✔ **Yes:**   *The farther the receiver stands away from the transmitter, the more transmission errors occur.*

2.7.28   header / heading

Use *header* when you mean the static area on top of each page.

Use *heading* when you mean the line of text that indicates what the subsequent section is about.

✔ **Yes:**   *In the header, insert the current chapter number and heading.*

✔ **Yes:**   *Make the font size of headings larger than the font size of body text.*

2.7.29   if / when / whether / whether or not

1 Use *if* to express a condition.

Use *when* for situations that require preparation. Also use *when* to denote the passage of time.

Use *whether* to express uncertainty and alternatives. Don't use *whether or not* in this case.

Only use *whether or not* if you want to emphasize that there are two possibilities or if you mean *under any circumstances*. However, it's often better to rephrase the sentence in this case.

2 Don't use *then* in if-clauses. It's a needless filler word and can even be misinterpreted in the sense that readers think that there's a time-based relation.

1

✔ **Yes:**   *If the engine starts ....*

   (means: It's not clear whether the engine will start.)

✔ **Yes:**   *When the engine starts ....*

   (means: At that point in time when the engine starts.

✔ **Yes:**   *If you need help, read the manual.*
✔ **Yes:**   *When Setup is complete, restart your computer.*
✔ **Yes:**   *To find out whether you need a new ink cartridge, ...*

✔ **Yes:**   *You must pay your taxes whether you want to or not.*
✔ **Top:**   *You must pay your taxes even if you don't want to.*

2

✘ **No:**   *If you need help, then read the manual.*
✔ **Yes:**   *If you need help, read the manual.*

Related rules

if / in case 231

If-clauses 232

2.7.30 if / in case

> Use *if* or *in case of* when talking about conditions.
>
> Use *in case* (without the word *of*) when talking about possibilities or precautionary actions.

✔ **Yes:**   *If it rains, take an umbrella.*

In case of rain, take an umbrella.

(means: Only take an umbrella if it rains; otherwise, don't.)

✔ **Yes:**   *Take an umbrella in case it rains.*

Take an umbrella in case there's rain.

(means: Take an umbrella as a precautionary measure.)

✔ **Yes:**   *Enter your phone number if you want us to call back.*

(means: Only enter your phone number if you want us to call you back; otherwise, don't.)

✔ **Yes:**   *Enter your phone number in case we need to contact you.*

(means: Enter your phone number now. We may need it if there's a problem.)

✔ **Yes:**   *If there's a fire, shut the door.*

✔ **Yes:**   *In case of fire, shut the door.*

Related rules

if / when / whether / whether or not [229]

If-clauses [232]

2.7.31   If-clauses

Many writers who speak English as a second language have trouble with if-clauses. In particular:

- Never use *would* in an if-clause.
- There's no comma before the word *if*.

Probable things in the future; cause and effect

main clause: **future or present**
if-clause: **present**

✔ **Yes:**   You'll be able to use the product if you follow the instructions.

✔ **Yes:**   Oil floats if you pour it on water.

✔ **Yes:**   If you don't wash your car, it gets dirty.

Less probable or unreal results of a condition that we imagine

main clause: **would**
if-clause: **past**

✔ **Yes:**   The tool would be helpful if you used it correctly.

✔ **Yes:**   If I were rich, I wouldn't spend my time writing manuals.

Mainly in British English also: *If I was rich, I wouldn't spend my time writing manuals.*

Impossible things in the past (the condition was not fulfilled)

main clause: **would have**
if-clause: **past perfect**

✔ **Yes:**   It would have worked better if you had chosen a smaller item.

✔ **Yes:**   If the sentence that had "had had" had had "had" it would have been correct.

Future actions that are unlikely

if + should + modal verb / imperative

✔ **Yes:**   *If the operation should fail, please contact support.*

Suggestions and hints

if + modal verb + modal verb

✔ **Yes:**   *If you can't find the key, it might be missing on your keyboard.*

Polite requests

if + will + modal verb

✔ **Yes:**   *If you will give us your address, we can send you the information.*

Related rules

if / when / whether / whether or not 229

if / in case 231

Commas 170

2.7.32   in / in to / into

> Use *in* when you want to indicate that something is within limits or within an area.
>
> Use *in to* when *in* is part of the verb.
>
> Use *into* to imply motion to the inside or interior of something (also figurative).

✔ **Yes:**   *A word is in a paragraph, but you copy the text into the document.*

✔ **Yes:**   *You log in to the computer.*

✔ **Yes:**   *Insert the DVD into the disc drive.*

Related rules

in / within 235

2.7.33   in / within

> Use *in* to refer to a location.
>
> Use *within* only when referring to a time frame or when you mean "inside the boundaries or limits of." Don't use *within* as an inflated version of *in*.

✔ **Yes:**   The name appears in the list.

✔ **Yes:**   Copy the file to a subfolder in the **Demo** folder.

✔ **Yes:**   Confirm your selection within five minutes.

▶ Related rules

in / in to / into 234

2.7.34   index / indexes / indices

When *index* means a list of words, like those in an alphabetical index, the plural is *indexes*.

The plural of a mathematical or physical *index* is *indices*.

✔ **Yes:**   *Choose one of the following indexes to look up a term: ...*

✔ **Yes:**   *Choose one of the following indices to measure the degree of pollution: ...*

2.7.35   it's not / it isn't

Both forms are correct contractions of *it is not*. It's perfectly OK to use either version.

When in doubt, use *it's not*. With *it's not*, the word *not* remains fully visible, which puts a stronger emphasis on the negation.

Tip:
Often, the best solution is to use a stronger verb or to transform a negative sentence into a positive one (see *Use strong verbs* 132 and *Be positive* 35).

✔ **Yes:**   *If it isn't cold, check the level of the cooling liquid.*
✔ **Top:**   *If it's not cold, check the level of the cooling liquid.*

✔ **Yes:**   *It isn't necessary to rewrite the whole text.*
✔ **Yes:**   *It's not necessary to rewrite the whole text.*
✔ **Yes:**   *You don't have to rewrite the whole text.*
✔ **Top:**   *You only need to rewrite small portions of the text.*

Related rules

Use contractions 125

2.7.36 lay / lie / lie

Don't confuse the words *lie* and *lay* in the various tenses.

You can *lay* things, but you cannot *lie* things.

- to *lay*, *laid*, *laid* means to *put* or *place*
- to *lie*, *lay*, *lain* means to *rest*
- to *lie*, *lied*, *lied* means to *deliberately tell something untrue*

✔ **Yes:** Lay the wrench down on the workbench.

✔ **Yes:** The cable lies on the floor.

✔ **Yes:** Don't lie to your readers.

2.7.37   less than / fewer than / under / below

Use *less than* when referring to quantities, figures, and amounts that relate to something that can't be counted or that doesn't have a plural. *Less* is also used with numbers when they are on their own and with expressions of measurement or time.

Use *fewer than* when referring to quantities, figures, and amounts that relate to discretely quantifiable nouns.

Use *under* only when referring to positions, especially if you want to convey the sense of *covering* or *crossing*.

Use *below* when referring to a lower-level position in general. To avoid ambiguity, you can also say *directly below* or *right below* to make clear that two things are in a vertical line and in close proximity.

✖ **No:**   The device consumes under 100 watts.

✔ **Yes:**   The device consumes less than 100 watts.

✔ **Yes:**   The new model consumes less energy than the old one.

✔ **Yes:**   The new model needs fewer battery cells then the old one.

✔ **Yes:**   The socket is hidden under the cover.

✔ **Yes:**   Move the specimen back and forth under the detector.

✔ **Yes:**   You can find the reset button on the backside directly below the power button.

✔ **Yes:**   Always place a figure heading below the figure, never above the figure.

Related rules

more than / over / above 243

2.7.38   let / leave

Let simply means *to allow.*

Leave means *to allow to remain.*

✔ **Yes:**   Don't let children play near the grill.

✔ **Yes:**   Don't leave children unattended.

✔ **Yes:**   Don't let the cooling liquid touch your skin.

✔ **Yes:**   Don't leave the cooling liquid in the engine for more than 6 months.

2.7.39 like / such as / as

In everyday speech, *like* and *such as* are often used interchangeably. In technical documentation, however, be precise:

- *like* means *similar to*
- *such as* means *as for example*

Don't use *like* as a conjunction; use *as* instead.

✔ Yes: *Office suites like LibreOffice provide a powerful word processor.*

(means: Only those office suites provide a powerful word processor that are similar to LibreOffice. There may be other office suites that are different and that don't provide a powerful word processor.)

✔ Yes: *Office suites, such as LibreOffice, provide a powerful word processor.*

(means: All office suites provide a powerful word processor. LibreOffice is an example of an office suite.)

✔ Yes: *Moving a dialog box is like moving a window.*

✘ No: *You can work with remote files like you would with local files.*

✔ Yes: *You can work with remote files as you would with local files.*

2.7.40   like / as if / as though

> Use the word *like* only before nouns, but never use the word *like* before a clause.
>
> If a subject and verb follow, use *as if* or *as though*. The meaning of *as if* and *as though* is identical.

✔ **Yes:**   *The spy camera looks like a pen.*

✔ **Yes:**   *The spy camera looks as if it was made for James Bond.*
The spy camera looks as though it was made for James Bond.

✔ **Yes:**   *The new model looks like the old one.*

✔ **Yes:**   *The new model looks as though it might break more easily than the old one.*
The new model looks as if it might break more easily than the old one.

✔ **Yes:**   *It appears as though you don't have permission to ....*

2.7.41   more than / over / above

Use *more than* when referring to quantities, figures, and amounts.

Use *over* only when referring to positions, especially if you want to convey the sense of *covering* or *crossing*.

Use *above* when referring to a higher-level position in general. To avoid ambiguity, you can also say *directly above* or *right above* to make it clear that two things are in a vertical line and in close proximity.

✖ **No:**   The device can store over 200 pictures.

✔ **Yes:**   The device can store more than 200 pictures.

✔ **Yes:**   The newly inserted picture appears over the background.

✔ **Yes:**   Move the detector slowly over the specimen.

✔ **Yes:**   You can find the reset button on the backside directly above the power button.

✔ **Yes:**   Always place a table heading above the table, never below the table.

Related rules

less than / fewer than / under / below `239`

2.7.42　many / much / a lot of / lots of

Use *many* only when referring to countable objects. Use *much* only when referring to abstract, inseparable, continuous, or otherwise uncountable quantities.

In positive statements, you can often also use *a lot of* (or *lots of*) for both countable objects and uncountable quantities. *A lot of* (or *lots of*) is usually more common in positive statements and more informal than *much*.

Much and *many* are more common in questions and negative statements.

✔ **Yes:**　*There's too much noise.*
✔ **Yes:**　*There are too many sounds.*

✔ **Yes:**　*There are many cars on the road.*
✔ **Yes:**　*There are a lot of / lots of cars on the road.*
✔ **Yes:**　*There was so much fuel in the tank that it spilled over.*
✔ **Yes:**　*There's a lot of / lots of fuel in the tank.*

✔ **Yes:**　*Are there many cars on the road?* (question)
✔ **Yes:**　*There aren't many cars on the road.* (negative statement)
✔ **Yes:**　*Is there much fuel left in the tank?* (question)
✔ **Yes:**　*There isn't much fuel left in the tank.* (negative statement)

Related rules

many / various / different 245

2.7.43 many / various / different

Use *many* when you mean *a large number of.*

Use *various* when you want to point out diversity.

Use *different* when you mean *distinct, separate,* or *unlike in nature, quality, form or degree.* Don't use *different* as a synonym for *many* or *various.*

✔ **Yes:** *Many users like our product.*

✔ **Yes:** *They love the various features of our product.*
✔ **Yes:** *In addition, they can choose from various colors.*

✔ **Yes:** *All colors are different even if some of them seem to be very similar.*
✔ **Yes:** *There are different user modes, such as easy mode, advanced mode, and expert mode.*

Related rules

many / much / a lot of / lots of 244

2.7.44   on / upon

> Use *on* rather than *upon*.
>
> Both words can be used interchangeably, but *on* is shorter and less formal.

✘ No: *This rule is based upon facts.*

✔ Yes: *This rule is based on facts.*

2.7.45   Possessives

Forming possessives from company names, product names, feature names, and objects looks and sounds odd because we're not used to these forms of the words.

Instead, use the name or object as an adjective, or use a construction with the word "of." (This is an exception to the rule *Watch for "the ... of" and for "of the"* 117.)

✘ No: *DemoCorporation's products*
✔ Yes: *DemoCorporation products*

or:

the products of DemoCorporation

✘ No: *DemoSoft's key features*
✔ Yes: *the key features of DemoSoft*

✘ No: *The button's name appears when you ....*
✔ Yes: *The name of the button appears when you ....*

✘ No: *Enter the disk's name.*
✔ Yes: *Enter the disk name.*

Related rules

Apostrophes 162

2.7.46   prior / previous

Prior refers to something that occurred earlier in time or with a higher order of importance.

Previous connotes *preceding* and always has a relation to a current object or event.

Don't use *prior* as a preposition to convey the idea of *before*. Also don't use *prior* to mean *beforehand* or *earlier*.

✔ **Yes:**   *If you have any prior measurement data, you can use this data instead of performing a new measurement.*

✔ **Yes:**   *The results of this measurement are higher than those of the previous measurement.*

✘ **No:**   *Calibrate the device prior to performing a new measurement.*

✔ **Yes:**   *Calibrate the device before you perform a new measurement.*

2.7.47   regular / standard / default / preset

Don't use *regular* when you mean *standard* (normal).

Use *regular* only when you mean symmetrically arranged or occurring at fixed intervals.

When talking about preconfigured settings:

- in user documentation, use *preset*
- in developer documentation, use *default*

✔ **Yes:**   *Use this setting for standard measurements.*

✔ **Yes:**   *This function draws a regular polygon.*

✔ **Yes:**   *The diagram shows some regular patterns at 5, 10, and 15 kHz.*

✔ **Yes:**   *If you hear a regular sound about once every 5 seconds, clean the air filter.*

✔ **Yes:**   In user documentation: *If you use the presets, you don't have to enter any parameters. The preset color is blue.*

✔ **Yes:**   In documentation for developers: *If you use the default settings, you don't have to enter any parameters. The default color is blue.*

2.7.48 safety / security

Use *safety* when referring to the protection from physical damage that's caused by natural disasters or accidents.

Use *security* when referring to the protection from damage that's caused by humans (often intentionally). This can be physical and non-physical damage.

✔ **Yes:** *Online banking is perfectly safe, but it's not always secure.*

✔ **Yes:** *job safety*

(Refers to the risk of being injured at the work place.)

✔ **Yes:** *job security*

(Refers to the risk of becoming unemployed.)

✔ **Yes:** *safety measures*

(Means, for example, the obligation to wear a helmet and protective goggles.)

✔ **Yes:** *security measures*

(Means, for example, measures against burglary and industrial espionage.)

2.7.49   single / singular

Single means *one* or *individual*.

Singular means *unique, unusual, distinctive, composed of one member*, or *composed of one kind*.

✔ **Yes:**   *There's a single button on the device.*

(means: There's no other button. There's only one.)

✔ **Yes:**   *There's a singular button on the device.*

(means: There's a button that's unique. You probably haven't seen a button like this before. This button may be the only button or there may be other buttons as well.)

2.7.50   that / which

Don't use *that* and *which* interchangeably or to avoid word repetitions. Word repetitions are perfectly OK in technical writing (see *Always use the same terms* [123]).

- Use *that* for restrictive clauses (clauses that cannot be removed without distorting the meaning).
- Use *which* for nonrestrictive clauses (clauses that can be put in parentheses or removed entirely).

Put a comma before *which* but don't put a comma before *that*.

✔ **Yes:**   *Press the key that's labeled Ignition.*

✔ **Yes:**   *Press the Ignition key, which launches the rocket.*

✘ **No:**   *Delete the line that defines the command which you want to remove.*

✔ **Yes:**   *Delete the line that defines the command that you want to remove.*

Note that *that* or *which* can imply a very different meaning:

✔ **Yes:**   *We accepted the last bid, which was sent by fax.*

   (means: We accepted the last bid. By the way, this bid was sent by fax.)

✔ **Yes:**   *We accepted the last bid that was sent by fax.*

   (means: We accepted the last bid of those bids that were sent by fax. However, there might have been other bids that we received later by mail or by email.)

✔ **Yes:**   *The file, which stores the passwords, was deleted.*

   (means: The file was deleted. Incidentally, this file stores the passwords.)

✔ **Yes:**   *The file that stores the passwords was deleted.*

   (means: A particular file was deleted. It was the password file.)

Related rules

Commas [170]

that / who / whom [254]

2.7.51   that / who / whom

> Use *that* and *which* to refer to things or to groups of people (see *that / which* [252]).
>
> Use *who* to refer to a single person.
>
> When used as an object, *who* becomes *whom*.
>
> Tip:
> If you can substitute *he*, *she*, *we*, or *they* in the clause, *who* is the correct word to use. If you can substitute *him*, *her*, *us*, or *them*, *whom* is the correct word.

✘ No: *John is the person that bought the device.*
✔ Yes: **John is the person who bought the device.**
✔ Yes: **John, who works for ABC, bought the device.**
✔ Yes: **This is the device that John bought.**

✘ No: *This is the team who designed the device.*
✔ Yes: **This is the team that designed the device.**
       (often better: **This team designed the device.**)

✘ No: *This is the virus who deleted the hard disk.*
✔ Yes: **This is the virus that deleted the hard disk.**
       (often better: **This virus deleted the hard disk.**)

✔ Yes: **Susan is the person for whom John bought the device.**
       (often better: **John bought the device for Susan.**)

Related rules
that / which [252]
Commas [170]

2.7.52   this / that / these / those

Use *this* (singular) and *these* (plural) if something is close in terms of place or time. Typically, *this/these* is used for something that's still in progress or in the future.

Use *that* (singular) and *those* (plural) if something is further away in terms of place or time. Typically, *that/those* is used for something that has already ended.

> ❶ **Important:** When using *this*, *these*, *that*, and *those*, make sure that it's clear which object you're referring to. If this isn't clear, add a syntactic cue. See *Be clear about what you're referring to* 106 and *Add syntactic cues* 104.

✔ **Yes:**   *Press this button, not that one.*

✔ **Yes:**   *Buy these items, not those from our competitor.*

✔ **Yes:**   *This is the correct way to do it, while that other way was wrong.*

✔ **Yes:**   *This is how a car looks today. That's how cars looked 50 years ago.*

✔ **Yes:**   *This is how you should proceed.*

Related rules

that / which 252

that / who / whom 254

Be clear about what you're referring to 106

2.7.53 use / utilize / employ

You *use* items when you use them the way they were intended to be used.

You *utilize* items when you use them in a way that they weren't designed for.

Employ suggests the use of a person or thing that was previously idle or inactive. To avoid confusion, especially with readers who speak English as a second language, don't use *employ* as a synonym for *use* or *utilize*. Use *employ* only when you mean *to hire*.

✘ **No:**   Utilize a screwdriver to tighten the screw.

✔ **Yes:**   Use a screwdriver to tighten the screw.

✔ **Yes:**   Utilize a screwdriver as a chisel.

✘ **No:**   Employ a colleague to help you.

✔ **Yes:**   Ask a colleague to help you.

✔ **Yes:**   The boss employed two additional experts.

2.7.54   what / which

Don't use *what* and *which* interchangeably.

What identifies: When there is a very large or unlimited number of possibilities that could be the answer, use *what*.

Which selects: When there is a limited number of possibilities, and when you're asking the reader to choose, use *which*.

With people, the number of people can always be considered limited, so you can always say *which*. However, *who* or *whom* is often shorter and clearer.

✔ **Yes:**   *What kind of car do you prefer?*

✔ **Yes:**   *I have a blue car and a red car. Which one do you prefer?*

✔ **Yes:**   *What you should enter depends on your personal goals.*

✔ **Yes:**   *Select which file format to use.*

✔ **Yes:**   *If you don't know which person to ask, read the manual.*

✔ **Top:**   *If you don't know whom to ask, read the manual.*

2.7.55   while / as / whereas / although

To avoid ambiguity, don't use the words *while*, *as*, *whereas*, and *although* interchangeably.

While means *at the same time*. Don't use *as*.

Whereas means *on the other hand*, or *on the contrary*.

Although means *in spite of the fact that*.

✘ **No:**   Keep these guidelines in mind as you write.
✔ **Yes:**   Keep these guidelines in mind while you write.

✘ **No:**   These guidelines are helpful, while others are not.
✔ **Yes:**   These guidelines are helpful, whereas others are not.

✘ **No:**   While the price is low, the product is excellent.
✔ **Yes:**   Although the price is low, the product is excellent.
✔ **Yes:**   While the price was low, we sold more pieces.

Related rules

because / since / as 205

2.7.56   want / wish / desire / need

Don't use *wish*, *want*, or *desire* when you mean *need*.

Use *want* rather than *need* when the readers' actions are optional.

Usually, use *want* rather than *wish* because *wish* is usually too strong and too emotional for technical communication. Use *wish* instead of *want* only if the word stands by itself at the end of a phrase or sentence.

Avoid the word *desire* completely because it's even stronger than *wish*.

✔ **Yes:**   Select the spare parts that you need.
✔ **Yes:**   Select the movies that you want to download.

✔ **Yes:**   Save the file if you want to keep it.
✔ **Yes:**   Rename the file if you wish.

✖ **No:**   Select the desired color.
✔ **Yes:**   Select the color that you want to apply.

✖ **No:**   Make the desired changes.
✔ **Yes:**   Make your changes.

2.8   FAQ: Standard terms and phrases

Use standard terms and phrases whenever possible. Standard terms and phrases help readers to identify objects and tasks quickly and free of doubt, which makes your documents easy to read.

In addition, if your documents are translated into foreign languages, using standard terms and phrases increases translation quality and minimizes translation costs.

What the recommendations are based on

The standard terms and phrases presented in this guide are based on:

- the rules given in the sections *Writing in general* 15, *Writing sections* 51, *Writing sentences* 87, and *Writing words* 113
- conventions of major software manufacturers (sometimes these companies prefer different terms; we've chosen the more common and the more user-friendly alternative in each case)
- international standards
- frequency of use of particular terms

> **Important:** If you have a good reason to use alternative terms, phrases, or spellings, feel free to do so. However, be consistent in the way you use them (see *Always use the same terms* 125).

Diversified style vs. unified style

There are two basic philosophies:

- You can use a unique term and a unique verb for each control and action (**diversified style**).
  - Advantage: prevents failure because the used terms clearly communicate where to act and what to do
  - Disadvantages: more complex terminology; longer texts
- You can use as few terms and verbs as possible (**unified style**).
  - Advantages: simplified terminology; slightly easier to translate
  - Disadvantage: readers need more time to identify the correct user interface control and action

A mixture of both styles is a **shortened diversified style**. Here, either the verbs *or* the names of controls are unique, but the other one is unified or omitted.

Example of diversified style:

1. From the *File* menu, choose *Print.*
2. Activate the *Printer Settings* tab.
3. Select the option *Landscape.*

Example of shortened diversified style:

1. Choose *File* > *Print.*
2. Activate *Printer Settings.*
3. Select *Landscape.*

or:

1. Click the menu item *File* > *Print.*
2. Click the tab *Printer Settings.*
3. Click the option *Landscape.*

Example of unified style:

1. Click *File* > *Print.*
2. Click *Printer Settings.*
3. Click *Landscape.*

Tip:
Use unified style only in very short documents or if you know that a document will be read by users who have very limited reading or language skills.

ℹ **Important:** This guide suggests a diversified style where diversification increases understandability, and it suggests a unified style where unification is unambiguous. If you prefer a completely unified style over the suggested mixture, use *click* instead of more precise verbs such as *select*, *choose*, and *activate*. In addition, for fully unified style, omit all terms that relate to specific user interface controls (see examples above).

Related rules

Be consistent 29

Always use the same terms 123

FAQ: Grammar and word choice 197

2.8.1 Hardware

Use the following terms and phrases. Also, note the given spellings.

Computer, add-on

Programs run on a *computer*.

The computer, together with all peripheral devices, accessories, and programs, is the *system*. Don't use *system* to refer to the computer alone. An *add-on* is an optional hardware device that's attached to the computer.

Don't use: *PC*, *machine*, *client*, *workstation*, *unit*

Board, card, adapter

Don't confuse *board* with *card*. A *board* is built in; a *card* can be removed by the user.

In general, try to avoid the term *card*. Use *adapter* to describe hardware that connects a network or a device to a computer. (Reason: Often, adapters can either be cards, or they can be integrated on the main board.) The term *card*, however, is OK when you make a specific reference to a device that has "card" in its name, such as *smart card*.

Typical adapters are: *graphics adapter*, *sound adapter*, *network adapter*

Don't use: *graphics card*, *sound card*, *network card*

Miscellaneous terms

Some frequently used terms are:

- *handheld* (one word, no hyphen)
- *touchscreen* (one word)
- *touchpad* (one word)
- *main board* (two words; don't use: *motherboard*)
- *keyboard* (one word)
- *firmware* (one word)
- *power adapter* (don't use: *AC adapter*)
- *power button* (don't use: *power switch*)
- *peripheral device*
- *onboard* ... (one word)
- *indicator light* (avoid *LED* if you mean the function rather than the type of electronic component)

Connecting devices

You *connect* a device *to* a computer.

You *connect* computers *to* a network.

Cables are *connected to* a device. One device is *connected to* another device.

Don't use: *to attach*, *to cable*, *cabled*

A connector can be *plugged into* a socket, jack, slot, or port.

The part that you plug in is the *plug*.
Don't use: *male connector*

A *socket* is a receptacle with holes of any type.
Don't use: *female connector*

A *jack* is a small, round, 1-pin socket used in audio and video connections.

A *slot* is a long, thin socket that allows you to insert cards.

The term *port* refers to a location for passing data in and out of a device.

Describe connectors in such a way so that even users who don't know the correct terminology can identify them clearly. Consider adding a picture.

2.8.2 Software

Use the following terms and phrases. Also, note the given spellings.

Operating system

There are *Windows users*, *Unix users*, and *Mac users*. There are *programs for Windows*, *programs for Unix*, and *programs for Mac OS*.
Don't use: *users under Windows*, *Windows application*, *Windows-based application*, and so on

A computer *runs* an operating system.

✔ **Yes:** *On a computer that runs Windows, ...*

Software *runs on* an operating system.
Don't use: *under*, *within*

✖ **No:** *DemoSoft runs under Windows.*

✔ **Yes:** *DemoSoft runs on Windows.*

✔ **Yes:** *In Windows, you can ....*

Program, software, application

Use *software* if you mean software in general.

Use *program* if you mean an individual program.

Use *application* only in material written for developers when you want to emphasize that you mean all components of a product, not just an executable file.

Use *web application* when you want to emphasize that a program runs in a browser.

Don't use: *executable*, *application program*, *software program*

> ℹ **Important:** Materials written for software on Mac OS often use the term *program* only for software that doesn't have a graphical user interface. In contrast, *application* is used to refer to all software that does have a graphical user interface.

In general, try to refer to a product by its name rather than by the vaguer term *program*.

✖ **No:** *The program stores your pictures.*

✔ **Yes:** *DemoSoft stores your pictures.*

There's *commercial software*, *freeware* (one word), and *open source software* (*open source* two words, lowercase).

Driver, software

In user documentation, avoid the term *driver*. Use *software*.

✘ **No:**   *Install the printer driver.*

✔ **Yes:**   *Install the printer software.*

Utility, tool, toolkit

Use *utility* if you mean an individual program. (Usually, a utility is a small, helpful program that's designed to perform a specific task rather than a broad range of tasks.)

Use *tool* if you mean a feature within a program.

Use *toolkit* if you mean a set of predefined routines rather than a *program* or *utility* as a whole.

Don't use: *utility program*, *utility application*, *toolbox*

Plug-in, add-in

Use *plug-in* or *add-in* (both with a hyphen) when referring to software that adds functionality to another (usually larger) program.

Use one of both terms consistently. When in doubt, use *plug-in* rather than *add-in*.

If one of both terms is part of the product's name, use this term. For example, if your product is advertised as "DemoSoft Power Plug-in," call it a *plug-in* and avoid the term *add-in* completely.

Note:
Don't confuse *add-in* with *add-on*. An *add-on* isn't software but *hardware* that's attached to a computer.

Widget

The term *widget* refers to an object that's changeable by the user.

In user documentation, when possible use a more general, non-technical term, such as *tool*.

Applet

The term *applet* refers to an HTML-based program that a web browser downloads temporarily.

In user documentation, when possible use a more general, non-technical term, such as *tool* or *program*.

2.8.3 Versions, updates

Use the following terms and phrases. Also, note the given spellings.

When to use version numbers

Don't mention a particular version number if the number doesn't add any important information. (Usually, it only does when you're talking about the differences between two versions.) Only specify the version number on the first mention in a topic or section.

If you must mention a version number, only make the version number as specific as needed.

Examples:

- When possible, don't mention the version number at all:
  "With *DemoSoft* you can ...."
  (The version isn't mentioned, so this statement refers to the latest version or to the version that's described in the document.)

- When needed, add the full version number:
  "With *DemoSoft version 5.2* you can ...."
  (You want to point out the difference from another minor version, most likely version 5.1 or version 5.0.)

- If the minor version isn't relevant, only mention the major version number:
  "With *DemoSoft version 5* you can ...."
  (You want to point out the difference from another major version, most likely version 4.)

Except in a dedicated "What's New" topic, avoid describing a product or feature as *new*. With the next version, this statement becomes obsolete. If it's important to point out that a particular feature hasn't been available in earlier versions, add the version number.

✖ No: *The new printing feature ....*

✔ Yes: *The printing feature, introduced in version 3.2, ...*

Version numbers

Use lowercase for the word *version*.
Don't use: *release*

It's OK to leave out the word *version*. Don't abbreviate it with the letter *v*.

Don't use *x* to mean *any number*.

✖ No: *In the 5th release of DemoSoft, ...*

✖ No: *In DemoSoft Version 5, ...*

✔ **Yes:** *In DemoSoft version 5, ...*
✔ **Top:** *In DemoSoft 5, ...*

✘ **No:** *DemoSoft works with OtherSoft v5.x.*
✔ **Yes:** *DemoSoft works with OtherSoft 5.*

In the middle of a sentence, don't add a period after the last number.
✔ **Yes:** *DemoSoft 5.0.1 is the current version.*
✔ **Yes:** *The current version is DemoSoft 5.0.1.*

Avoid the word *running* to refer to a specific version. Instead, use the words *use*, *using*, or *installed*.
✘ **No:** *If you're running DemoSoft 3 or an earlier version, ...*
✔ **Yes:** *If you're using DemoSoft 3 or an earlier version, ...*
✔ **Yes:** *To use this feature, you must have DemoSoft 4 or a later version installed.*

Earlier versions, later versions

There may be *earlier* versions of a program and *later* versions.
Don't use: *prior, lower, higher, older, newer, above, below, smaller, bigger, greater, better, modern, legacy*
✔ **Yes:** *... requires DemoSoft version 5 or a later version.*
✔ **Yes:** *... requires DemoSoft version 5 or an earlier version.*
✔ **Yes:** *Make sure that you're using the latest version.*
✔ **Yes:** *To convert data from earlier versions, ...*

A program can be *backward compatible* and *forward compatible*.
Don't use: *future compatible, upward compatible*

Upgrades, updates

Upgrade refers to a new major version. *Update* refers to a new minor version.
✔ **Yes:** *If you want to upgrade DemoSoft 4, ...*

   (means: If you want to install version 5 or a later version, ...)
✔ **Yes:** *If you want to update DemoSoft 4, ...*

   (means: If you want to install version 4.x, ...)

Instead of *update*, use the term *critical update* for a bug fix that can affect the core functionality of your product but that isn't related to security. For

security-related bug fixes, use *security update*.
Don't use: *bug fix*, *patch*, *maintenance release*

Requirements

Software may be *compatible with* other software or with hardware. If it's compatible, it *works with* other software or hardware.
Don't use: *supports*

✖ **No:**   *DemoSoft supports the XYZ standard.*

✔ **Yes:**   *DemoSoft is compatible with XYZ standard.*

✔ **Yes:**   *DemoSoft works with XYZ.*

2.8.4 Setup

Use the following terms and phrases. Also, note the given spellings.

Installation, setup

Use *install* to refer to adding hardware or software to a computer system. You install items *on* a disk (not *onto* a disk).

Don't use *install* when you mean *installation*. It's the *installation disc*, the *installation procedure*, and the *installation wizard*.
Don't use: *install disc*, *install procedure*, *install wizard*

The software used for installation is the *installer*. Capitalize *installer* only when you refer to a specific installer application.
Don't use: *installation utility*

✔ **Yes:** *The DemoSoft Installer.*

✔ **Yes:** *Most programs come with an installer.*

Use *set up* (two words) to refer to the process of preparing and configuring a program. This procedure is the *setup* (one word).

Often, the basic setup is done by the installer. For this reason, it's usually adequate to call the program that performs both installation and setup the *installer* rather than the *setup program*.

Don't use *setup* as a short form of *setup program*.

Removing, uninstalling

In general, use the term to *remove*. Use *uninstall* only if the user interface also uses this term.

The software used for removing a program is the *uninstaller*.

Don't use: *deinstall*, *delete*, *deinstaller*

Talking about the registry

Don't capitalize the word *registry*.

To refer to the entries within the registry in general, use the term *registry settings*.

There are *registry keys*, *registry entries*, and *values*.
Don't use: *registry value*

Related rules

Versions, updates 268

2.8.5   Start, stop

Use the following terms and phrases. Also, note the given spellings.

Switching on the computer

In general, *switch on / switch off* is more common in Britain English, and *turn on / turn off* is more common in American English.
Don't use: *power on, power off, power up, power down*

Starting the system

You *start* or *restart* the system.
Don't use: *start up, boot, reboot*

Starting a program, opening files

You *start* a program, but you *open* an application (for different uses of the terms *program* and *application*, see *Software* 265).

You *open* a file.

You *access* data (if you need to obtain access to it), or you just *read* data.

Don't use: *call, call up, launch, invoke*

Running programs

You *run* a program or application.
Don't use: *execute*

Don't use *run* to describe what users do with a program.

✘ No: *Before installing the program, run the DemoConversion Utility to convert your existing data.*

✔ Yes: *Before installing the program, use the DemoConversion Utility to convert your existing data.*

Don't use *running* to refer to an *open* application.

✘ No: *If any application is running, ...*

✔ Yes: *If any application is open, ...*

When something happens while a program is running, it happens *at runtime* (runtime is one word).

If something is performed without any delay, it's done *in real time* (real time is two words).

Command line

You can also run many programs *from the command line*. These programs have a *command-line interface*.

At the command prompt, enter the program name along with the required *command-line parameters*.

Don't use: *on the command line, command-line prompt, system prompt*

Logging in and out

You *log in to* a program or web site. When you're finished, you *log out*.

You *log on to* a computer or network. When you're finished, you *log off from* it.

Don't use: *log into, log out of, log out from, quit, sign in, sign out*

The corresponding processes are the *login* and the *logout* (both one word, no hyphen).

There may be a *login window* and a *logout button*.

To log in, you need a *user name* and a *password*.

User rights

In Windows, there are *standard users* and *administrators*.
Don't use: *normal user, admin*

Different users may have different *user rights*, depending on their *user accounts*.
Don't use: *access rights, access privileges, user privileges, privileges*

Don't confuse *user rights* with *permissions*.

User rights apply to system operations (for example, the right of an administrator to add new users).

Permissions apply to specific shared system resources (for example, the permission to write to a file while it isn't locked by another user).

Quitting applications

Normally, you *exit* a program. You *quit* an application (for different uses of the terms *program* and *application*, see *Software* ⌐265⌐).

If the program doesn't respond, you need to *close* it.

You *close* a document.

You *close* a window.

You *close* a file.

You *stop* hardware operations.

You *end* communications and network connections.

Don't use: *abort, leave, terminate*

Shutting down the system

You *shut down* the system before you *switch off* the computer. The process of doing so is the *shutdown*.

2.8.6 Assistants, documentation

Use the following terms and phrases. Also, note the given spellings.

Assistants, wizards

You can use either *assistant* or *wizard*. When in doubt, use the term *assistant* because it's less idiomatic.

Capitalize the words *assistant* and *wizard* only when the word is part of a name.

✔ **Yes:**   *Use the Print Assistant to ....*

✔ **Yes:**   *Use the assistant to ....*

An assistant or wizard consists of a sequence of *pages*.

✔ **Yes:**   *On the first page of the wizard, ...*

Tips

When the user holds the pointer over a button or other control, a *tip* may appear.
Don't use: *tooltip, infotip*

✔ **Yes:**   *The tip explains ....*

Help

In general, use the short form *help* rather than the lengthy form *online help*.
Use *online help* only if you want to point out the difference to a printed manual.

Capitalize only when referring to the complete name of a program's help.

✘ **No:**   *See online help for more information.*

✔ **Yes:**   *See help for more information.*

✔ **Yes:**   *See DemoSoft Help for more information.*

The button that users can press to open help is the *Help button*.
Don't use: *?, question mark, ?-button, question-mark button*

Don't use *how-to* as a noun.

✘ **No:**   *Help provides how-tos on ....*

✔ **Yes:**   *Help provides how-to information on ....*

Manuals

In general, avoid the term *manual* as a synonym for *book*, *guide*, or other specific terms referring to product documentation.

When possible, use the title of the book itself.

✖ **No:**   See the manual for more information.

✔ **Yes:**   See the DemoSoft User's Guide for more information.

Chapters, topics

Only use the term *chapter* in a printed manual. In online help, use *topic*.

When generating printed manuals and online help from the same text base (single source publishing), use *topic*, which is the more neutral term. *Topic* is acceptable for both printed manuals and online help. However, often the best solution is to avoid the terms *chapter* and *topic* altogether.
Don't use: *entry*, *article*

✔ **Yes:**   The topic "Customizing the layout" provides further information.

✔ **Top:**   For more information, see "Customizing the layout."

Screen shots, screen captures

Avoid the terms *screen shot*, *screen capture* , and *screen dump* in user documentation. Use *picture*.

Tip:
If you need to explicitly mention a particular picture within your text, this is often an indicator that you should restructure the topic. Position the picture closer to the text that talks about the picture. Consider adding visual clues or callouts to the picture that point out the elements that you're talking about. Then, you don't have to mention the picture in your text at all because it's evident what you mean.

✖ **No:**   As you can see in the screen shot, the text is now bold.

✔ **Yes:**   As you can see in the picture, the text is now bold.

✔ **Top:**   The text is now bold. (On the picture, add an arrow that points to the bold text, or add a callout that says, "Text is now bold.")

2.8.7   Windows, messages

Use the following terms and phrases. Also, note the given spellings.

Display, screen

Use *display* if you mean the complete physical device, but use *screen* if you mean the area where programs are shown.

Don't use *monitor* because many devices have a built-in display instead of an external monitor.

✔ **Yes:**   *Adjust the position of your display so that the top of the screen is slightly below eye level.*

✔ **Yes:**   *A window appears on the screen.*

Desktop

Use: *on the desktop* (*desktop* is one word, lowercase)
Don't use: *workspace*

✔ **Yes:**   *A new icon appears on the desktop.*

✔ **Yes:**   *Double-click the DemoSoft icon on the desktop.*

Startup screen

Use: *startup screen*
Don't use: *splash screen*, *startup display*, *opening display*

✔ **Yes:**   *If you don't want to see the startup screen when the program starts, ….*

Windows in general

A window *appears*.
Don't use: *opens*, *pops up*, *shows up*, *displays*

However, when possible, avoid stating that an item appears because users can see this anyway. Often, it's better to include an item's name in the description of the subsequent step of a procedure. (This doesn't break the rule of not merging two steps into one sentence because it only specifies the place where to act and is not truly a full action by itself.)

A program may *display* a message or document.

✖ No: *4. The Edit window displays.*

5. Select the option Layout.

✔ Yes: *4. The **Edit** window appears.*

*5. Select the option **Layout**.*

✔ Top: *4. In the **Edit** window, select the option **Layout**.*

✔ Yes: *If the file already exists, DemoSoft displays a warning message.*

✔ Yes: *DemoSoft now displays the picture that you've selected.*

When you refer to a window by name, use the exact words and capitalization as given in the title bar of the window. Don't capitalize the words *window* and *dialog*.

✖ No: *The editing window appears.*

✔ Yes: *The **Edit** window appears.*

You *open* and *close* windows.

You *switch between* open windows. You *switch to* a window or document. Don't use: *activate*

✖ No: *Activate the Edit window.*

✔ Yes: *Open the **Edit** window.*

✔ Yes: *Switch to the **Edit** window.*

On top of a window is the *title bar*.

On the right side of the title bar, windows have a *close button*, a *maximize button*, a *minimize button*, and a *restore button*.

When you click the program icon on the left side of the title bar, the *control menu* appears.

The window on top that you're currently using is the *active window*. Don't use: *current window*, *window that has the focus*

Something is *on the user interface* but *in a window*.

Dialogs and property sheets

A *dialog* is a window that requests information from the user. It must be explicitly dismissed by clicking a button.

In user documentation, also handle property sheets like dialogs and call them *dialogs*.

In user documentation, use the term *message* if the primary content of a dialog is a message (see the section on messages below).

Don't use: *dialog box*, *dialog window*, *pop-up window*, *property sheet*

When referring to a dialog by its name, use the exact words and capitalization as given in the title bar of the window.

A dialog is a special kind of window. For this reason, similar to windows, dialogs *appear*.
Don't use: *displays*, *opens*, *shows up*, *pops up*

✘ No: *The Print dialog box displays.*

✔ Yes: *The Print dialog appears.*

Messages and confirmations

In user documentation, try to refer to all types of messages as just a *message*. Use the verb to make a distinction between *warning messages* and *error messages*. Use *to inform* for confirmation messages and for error messages, and use *to warn* for warning messages.

Similar to windows and dialogs, messages *appear*. However, when possible, use a sentence with either *informs* or *warns*.

✔ Yes: *A message informs you that ….*

✔ Yes: *A message warns you that ….*

When you need to refer to a particular message:

- With a confirmation message, refer to its title.
  If the title only consists of a program name, refer to the main instruction in the message.
- With an error message, refer to the main instruction in the message.
  If the main instruction is too long, summarize it.
- With a warning message, refer to the question in the message.
  If the warning message doesn't ask a question, refer to the main instruction.

✔ Yes: *In the Report printed message, click OK.*

✔ Yes: *If an Insert new disc message appears, remove … and insert ….*

✔ Yes: *In the Replace existing file? message, click Yes.*

Notifications

Notifications appear in the *notification area* on the right side of the Windows taskbar.
Don't use: *alert*, *pop-up message*

✔ Yes: *When the "New demo device found" notification appears, click the notification to start the configuration.*

Panes, split bars

Use *pane* only to refer to distinct areas in a window. Because panes aren't labeled, use lowercase when referring to a particular pane.

✔ **Yes:**   *On the property pane, you can see ....*

The horizontal or vertical double line that separates a window into two panes is a *split bar*.

If a split bar isn't visible by default, there's a *split box* at the top right of the vertical scroll bar (for horizontal splitting) or at the far right of the horizontal scroll bar (for vertical splitting). Users can point to the split box and then *drag* the split bar.

Related rules

Controls [282]

Mouse, touchscreen [302]

Scrolling, zooming [306]

Deleting, cutting, pasting [309]

Keys [310]

2.8.8 Controls: General rules

When writing the steps of procedures, you often need to refer to a particular user interface control and to tell the users exactly what to do.

Watch the order

Make it easy for the readers to follow your instructions step by step. Match the order of words with the sequence of steps that users must take to identify an object: When possible, first name the control, and then tell the reader what to do with the control.

✖ No: *Click Save in the Options window.*

✔ Yes: *In the **Options** window, click **Save**.*
(Readers must first find the "Options" window, and then find and click the "Save" button.)

Referring to a control

To refer to a particular control, quote its exact label text (including capitalization). If there are three periods at the end of a menu item's name (an ellipsis), don't include these periods.

✔ Yes: *From the menu, choose **File** > **Save As**.*

If an interface element doesn't have any label, find an appropriate name yourself, and use this name consistently. Capitalize the name. For example, if there's a button with a symbol that shows a magnifying glass, you could call it the "Search" button.

✔ Yes: *Click the **Search** button.*

In user documentation, avoid using the full technical names of user interface controls. Instead, refer to an element by its caption and a more common name such as *list* instead of *drop-down list*.

✔ Yes: In user documentation:
*In the **Font size** list, select **Large fonts**.*

✔ Yes: In developer documentation:
Add a drop-down list box to the form.

Selecting, choosing

Use *select* to refer to the action that users perform when they select among multiple options or objects.
Don't use: *choose, activate, mark*

Exception: Use *choose* when referring to menu items (see *Controls: Menus – Ribbons – Toolbars* [285]).

✘ No: *Choose a name in the list of users.*

✔ Yes: *Select a name in the list of users.*

✔ Yes: *From the **File** menu, choose **Print.***

Also use *select* when users highlight text for editing. Selected text is *highlighted*.
Don't use: *mark*, *inverted*

✘ No: *Mark the text that you want to copy.*

✔ Yes: *Select the text that you want to copy.*

Things that aren't selected are *unselected*.
Don't use: *deselected*, *dehighlighted*

You *cancel a selection* in general. You *clear a check box* in particular.
Don't use: *deselect*, *unmark*

Enabling, disabling, activating

Use the terms *enable*, *enabled*, *disable*, and *disabled* only in developer documentation.

In user documentation, use *turn on*, *turn off* as general terms. If you're talking about a particular control, use the specific term for that particular control. For example, use *select* or *clear* when talking about a check box.

✘ No: *Disable your virus protection software.*

✔ Yes: *Turn off your virus protection software.*

✔ Yes: *Clear the **Make backup** check box.*

✘ No: *Make sure that the check box is enabled.*

✔ Yes: *Make sure that the check box is selected.*

Unavailable items

In user documentation, refer to unavailable items as *unavailable*. In developer documentation, use *disabled*.

To describe the appearance of an unavailable item, use *dimmed*.

Don't use: *grayed*, *shaded* (*shaded* is used when there's a mixture of settings for a selection in a group of options)

✔ Yes: *When working on a text document, image editing functions are unavailable.*

✔ Yes: *The command isn't available because no picture is selected.*

✔ **Yes:**   *Some commands may appear dimmed if they're unavailable in your version.*

Presets, defaults

In user documentation, use *preset*. In developer documentation, use *default*. See also *regular / standard / default / preset* [249].

Tip:
Often, the best solution is to avoid the words *preset* or *default* altogether.

✔ **Yes:**   *The preset format is HTML.*

✔ **Top:**   *If you don't select a format, the program saves as HTML.*

Related rules

Controls: Menus, Ribbons, Toolbars [285]

Controls: Tabs, Groups [288]

Controls: Buttons, Command links [290]

Controls: Check boxes, Options [292]

Controls: Fields, Boxes, Sliders [294]

Controls: Lists [296]

Controls: Taskbar, Status bars [300]

Windows, messages [278]

Mouse, touchscreen [302]

Scrolling, zooming [306]

Resizing, aligning [308]

Deleting, cutting, pasting [309]

2.8.9 Controls: Menus, Ribbons, Toolbars

Use the following terms and phrases. Also, note the given spellings.

> **ⓘ Important:** In addition to the rules listed here, see *Controls: General rules* [282].

Menus

Item 1	Item 2	Item 3	Item4
Item 1A			
Item 1B			
Item 1C			
Item 1D			
Item 1E			

Use: *menu, submenu, shortcut menu*
Don't use: *drop-down menu, pull-down menu, secondary menu, cascading menu, pop-up menu, context menu, right-click menu*

The round or square button in the upper left corner of a ribbon opens the *Application menu*. The *Start menu* on the *taskbar* is used for starting programs.

Use: *menu command, menu item* (*menu item* is mainly used for items that aren't commands, such as names of windows in the Window menu); usually, however, you don't have to use these terms at all; for example, just say: *from the File menu, choose Open.*
Don't use: *menu option, choice*

Use: *from ... choose, choose*
Don't use: *in the menu, open, select* (you *select* a check box, and you *select* an item in a list)

Always stick to the same order that users must follow through the menu hierarchy.

✖ **No:** *Choose Print from the File menu.*

✔ **Yes:** If you want to emphasize that users must look at the menu rather than at another control:

diversified style:

From the File menu, choose Print.
Choose the menu command View > Font Size > Large.

unified style:
On the File menu, click Print.
On the View menu, click Font Size, and then click Large.

✔ **Yes:**   If it's obvious that you mean a menu:
Choose File > Print.

Ribbon, toolbars

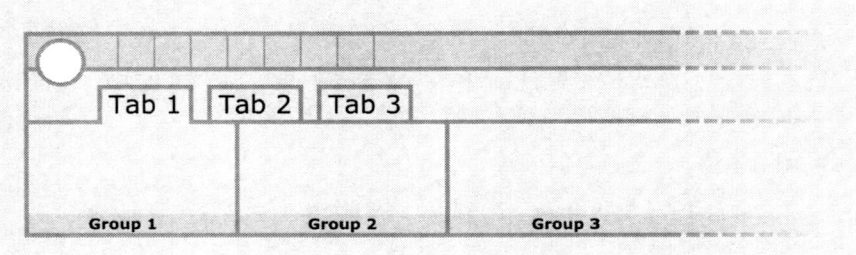

The *ribbon* consists of *tabs*, *groups*, and *controls*.

The round or square button in the upper left corner is the *Application button*. When you click this button, the *Application menu* appears. When possible, avoid these terms and use the name of the program instead. For example, if your program is called "DemoSoft," use *DemoSoft button* and *DemoSoft menu*.

The toolbar on the ribbon is the *Quick Access toolbar*.

With all toolbars, refer to the buttons and symbols on a toolbar by their labels. If a button or symbol has no label, use the text of the tip that appears when you hold the pointer over the control (for details, see the following section on icons and symbols).

Refer to menu buttons by their labels plus the word *menu*.

✔ **Yes:**   *On the Quick Access toolbar, on the Write tab, click Find & Replace.*

✔ **Yes:**   *On the Write tab, in the Font box, enter "Arial."*

✔ **Yes:**   *On the Write tab, in Paragraph Settings, select Center text.*

✔ **Yes:**   *On the toolbar, from the Insert Link menu, choose File Link.*

Icons, symbols

DemoSoft

Use the term *icon* only for program icons, which are usually either on the desktop or on the taskbar.

For user interface elements that are identified by a graphic rather than by a text label, use *button*. If it's inappropriate to call the icon a button, use *symbol*.

✔ **Yes:** *To start the program, double-click the DemoSoft icon.*

✔ **Yes:** *To find a particular term, click the* **Search** *button.*

✔ **Yes:** *Drag the document symbol to the list.*

If possible, avoid naming the control altogether (see *Controls: Buttons, Command links* 290).

✔ **Yes:** *To find a particular term, click the* **Search** *button.*

✔ **Yes:** *To find a particular term, click* **Search**.

> Related rules
>
> *Controls: General rules* 282
>
> *Windows, messages* 278
>
> *Mouse, touchscreen* 302
>
> *Scrolling, zooming* 306
>
> *Keys* 310

2.8.10 Controls: Tabs, Groups

Use the following terms and phrases. Also, note the given spellings.

> **ℹ Important:** In addition to the rules listed here, see *Controls: General rules* 282.

Tabs

Use: *tab*
Don't use: *tab card*, *index card*, *property sheet*

Use *click* plus the name of the tab. If you prefer diversified style, you can also use *activate*.
Don't use: *enable*, *select*, *open*, *mark*, *bring to front*

Often, the best solution is not to use any verb at all and to merge the sentence with the following sentence. (This doesn't break the rule of not merging two steps into one sentence because it only specifies the place where to act and is not truly a full action by itself.)

✔ **Yes:**   *1. Click the View tab.*
           2. Select ....

✔ **Yes:**   *1. Activate the View tab.*
           2. Select ....

✔ **Top:**   *1. On the View tab, select ....*

Group boxes

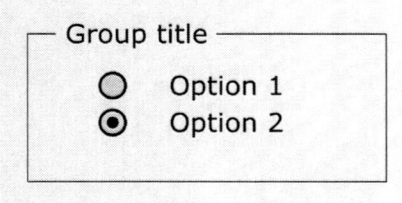

Only refer to the label of a group box; don't mention that it's a group box.
Don't use: *group*, *group box*

However, only mention the group box label if a dialog contains at least two options with identical names.

Use: *under*
Don't use: *in*

✔ **Yes:**   If there's only one option labeled "Hidden" in a dialog:
Select Hidden.

✔ **Yes:**   If there are two or more options labeled "Hidden" in the same dialog, and one of these options is in the group box "Effects":
Under Effects, select Hidden.

Progressive disclosure controls

Show all

If the control has a label, refer to the control by this label.

If the control doesn't have a label, refer to the control by its type, usually *plus/minus button*, *triangle*, or *arrow*.

Use: *click*
Don't use: *activate*, *open*, *expand*, *toggle*, *select*

✔ **Yes:**   *To view a list of all pictures, click Show all.*

✔ **Yes:**   *To reveal more information, click the triangle next to the heading.*

Related rules

Controls: General rules 282

Windows, messages 278

Mouse, touchscreen 302

Scrolling, zooming 306

Keys 310

2.8.11   Controls: Buttons, Command links

Use the following terms and phrases. Also, note the given spellings.

> **ⓘ Important:** In addition to the rules listed here, see *Controls: General rules* ⌐282⌐.

Buttons

Use: *button*
Don't use: *action button*, *push button*, *command button*, *key*

Use: *click*
Don't use: *activate* (use *activate* only for the action of verifying that a software product has a valid license), *press* (use *press* only for mechanical buttons such as the power button)

Include the word *button* only if it isn't obvious that you're talking about a button.

✔ **Yes:**   *To be able to edit the document, click the Modify button.*

✔ **Top:**   *To be able to edit the document, click Modify.*

Command links

Click to **trigger a function**

Don't mention that it's a link.

Use: *click*
Don't use: *follow*

✘ **No:**   *To copy the file, follow the link Duplicate.*

✔ **Yes:**   *To copy the file, click Duplicate.*

> **Related rules**
>
> *Controls: General rules* ⌐282⌐
> *Windows, messages* ⌐278⌐

2.8.12   Controls: Check boxes, Options

Use the following terms and phrases. Also, note the given spellings.

> ❶ **Important:** In addition to the rules listed here, see *Controls: General rules* 282.

Check boxes

☑ Setting 1
☑ Setting 2

Use: *check box* (two words)
Don't use: *box*, *option box*

Use: *select*, *clear*
Don't use: *activate*, *deactivate*, *mark*, *unmark*, *deselect*, *click* (this is ambiguous because you don't know the current state), *enable*, *disable*, *check*, *uncheck*, *choose* (you *choose* menu items)

Use *dimmed* when there's a mixture of settings for a selection in a group of options. The dimmed appearance indicates that some previously selected options affect the settings.

✔ **Yes:**   *Select the Demo check box.*

✔ **Yes:**   *Clear the Demo check box.*

Options

○ Option 1
◉ Option 2

Use: *option*
Don't use: *radio button*, *option button*

Use: *select*
Don't use: *click*, *check*

✔ **Yes:**   If you need to point out that you're talking about an option:
Select the option Print all pages.

✔ **Yes:**   If it's obvious that you're talking about an option:
Select Print all pages.

Related rules

2.8.13 Controls: Fields, Boxes, Sliders

Use the following terms and phrases. Also, note the given spellings.

> **ℹ Important:** In addition to the rules listed here, see *Controls: General rules* 282.

Standard fields

Your input: [|]

You can use either *field* or *box*. Use *field* for advanced users, and *box* for inexperienced users. Don't mix both terms within the same document.
Don't use: *entry field*, *input field*, *text box*

Use *type into* if typing or pasting is required. Use *enter* if typing or pasting isn't necessarily required, for example, if you can also select a value from a list or use a spin control. When referring to a form that has multiple text boxes, you can also use *fill in*.
Don't use: *to input*, *to key in*

✔ **Yes:** *Type your password into the **Password** box, and then click OK.*

✔ **Yes:** *Type your password into the **Password** field, and then click OK.*

✔ **Yes:** *Enter the date.* (When you can either type the date or select it from a calendar.)

✔ **Yes:** *Fill in the required fields, and then click **OK**.*

Boxes with an option to select from a list of entries

Your input: [| ▼]

Use: *box*
Don't use: *combo box, select box*

Use: *enter*
Don't use: *click, choose, select, type*

✔ **Yes:** *In the **Width** box, enter the paper width.*

Boxes with a spin control

Your input: 101

Use: *box*
Don't use: *spin control*

Use: *enter*
Don't use: *type, select*

If the text can't be edited via the keyboard, refer to the symbols of the spin control.

✔ **Yes:**   *In the **Year** box, enter your year of birth.*

✔ **Yes:**   *Click one of the arrows to increase or decrease the number of copies.*

Sliders

Use: *slider*
Don't use: *slide control, slider control*

Use: *move ... to the left, move ... to the right, move ... up, move ... down*
Don't use: *push, drag*

✔ **Yes:**   *To enlarge the object, move the **Size** slider to the right.*

Related rules

Controls: General rules 282

Windows, messages 278

Mouse, touchscreen 302

Scrolling, zooming 306

Keys 310

2.8.14 Controls: Lists

Use the following terms and phrases. Also, note the given spellings.

> **ⓘ Important:** In addition to the rules listed here, see *Controls: General rules* 282.

Hierarchical lists

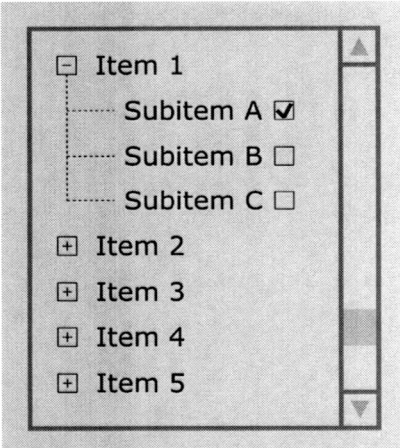

Use: *list*. You can also use *hierarchical list* if you want to highlight the fact that the list is organized hierarchically.
Don't use: *tree view, tree, list view*

Use: *in*
Don't use: *on, from*

Users *select* data, and *expand* and *collapse* list items.
Don't use: *choose, click*

If there are check boxes in the list, handle them like ordinary check boxes (see *Controls: Check boxes, Options* 292).

✔ **Yes:** *In the Contents list, select Writing help files.*

✔ **Yes:** *In the Files to back up list, clear the Demofiles check box.*

Tabular lists

Column 1	Column 2	Column 3	
List item A	List item A	List item A ☐	▲
List item B	List item B	List item B ☑	
List item C	List item C	List item C ☑	
List item D	List item D	List item D ☑	
List item E	List item E	List item E ☐	▼

Use: *list*
Don't use: *list view*, *list box*

Use: *in*
Don't use: *on*, *from*

Use *click* for the headings and *select* for the data.
Don't use: *choose*, *activate*

If there are check boxes in the list, handle them like ordinary check boxes (see *Controls: Check boxes, Options* 292).

✔ **Yes:** *In the Programs list, click DemoSoft, and then select Run at startup.*

✔ **Yes:** *In the Files to back up list, clear the Demofiles check box.*

Standard lists

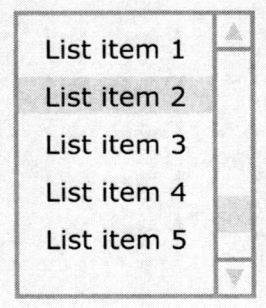

Use: *list*
Don't use: *list box*, *field*

Use: *in*
Don't use: *on*, *from*

Use: *select*
Don't use: *choose*, *click*, *activate*, *mark*

✔ **Yes:** *In the Font list, select Arial.*

Collapsible lists

Use: *list*
Don't use: *drop-down list*, *field*

Use: *in*
Don't use: *on*, *from*

Use: *select*
Don't use: *enter*, *choose*, *click*

✔ **Yes:** *In the Font style list, select Bold.*

Related rules

Controls: General rules 282

Windows, messages 278

2.8.15   Controls: Taskbar, Status bars

Use the following terms and phrases. Also, note the given spellings.

> **ⓘ Important:** In addition to the rules listed here, see *Controls: General rules* [282].

Taskbar

The Windows *taskbar* consists of the following items (from left to right): The *Start button* (which opens the *Start menu*), the *quick launch bar*, *taskbar buttons* for running programs, and the *notification area*.

Refer to all items on the taskbar as *buttons* and by their labels. If there's no label, use the text of the tip that appears when you hold the pointer over the control.

Status bar, status display

Use *status bar* (two words) if it consists of a single line of text. Use *status display* or *status area* if it's a larger area.
Don't use: *status line*

✔ **Yes:**   *You can see the current page number on the status bar.*

Progress indicators

A *progress indicator* shows the status of an action.
Don't use: *progress bar*

Related rules

Controls: General rules 282

Windows, messages 278

Mouse, touchscreen 302

Scrolling, zooming 306

Keys 310

2.8.16 Mouse, touchscreen

Use the following terms and phrases. Also, note the given spellings.

Input devices in general

A mouse, a trackball, a trackpad, a pen, or a finger on a touchscreen are *input devices*.

If you need to refer to more than one mouse, use *mouse devices*.
Don't use: *mice*

Never use *press* for items on screen. Use *press* only when referring to pressing keys on the keyboard, and when referring to pressing other mechanical buttons and switches.

Pointer, pointing, insertion point

Refer to the mouse pointer as the *pointer*.
Don't use: *mouse pointer, cursor, mouse cursor, I-beam, arrow*

Refer to the position where a new letter appears when you press a key on the keyboard as the *insertion point*.
Don't use: *cursor, caret, insertion mark*

The insertion point *blinks*.
Don't use: *flashes*

Avoid using the verbs *point, hover,* and *mouse over*. Instead, directly refer to the pointer, or use a more precise verb, such as *drag*.

✘ **No:**   *Point to the symbol to see a description.*

✘ **No:**   *Hover the mouse over the symbol to see a description.*

✔ **Yes:**   **Hold the pointer over the symbol to see a description.**

✘ **No:**   *Point to the button and drag it to the toolbar.*

✔ **Yes:**   **Drag the button to the toolbar.**

Mouse buttons

You *click* items on screen.

You *press* and *release* a mouse button.

Just say *hold down* a mouse button when you mean *press and then hold down*. (Reason: You can't hold it down without pressing it, so you don't need to say that it must be pressed.)
Don't use: *click and hold, press and click*

Just say *mouse button* when you mean the primary (usually left) mouse button. Use *right mouse button* for the secondary mouse button.
Don't use: *primary mouse button*, *secondary mouse button* (Reason: Users who have reprogrammed their mouse know that their mouse uses different buttons.)

Clicking

Users *click* things.
Don't use: *click on*, *click at*, *press*

✘ **No:** *Press the Save button.*

✘ **No:** *Click on the Save button.*

✔ **Yes:** If it's helpful to point out that you mean a button:

*Click the **Save** button.*

✔ **Yes:** If it's obvious that you mean a button:

*Click **Save**.*

Users don't click windows but *click in a window* or region.

✘ **No:** *Right-click the Outline window.*

✔ **Yes:** *Right-click anywhere in the **Outline** window.*

✔ **Yes:** *Click the title bar of the **Demo** window.*

Just say *to click* when you mean to click with the primary (usually left) mouse button.
Don't use: *left-click*

Say *to right-click* when you mean to click with the secondary (usually right) mouse button.

Say *to double-click* when you mean two clicks immediately one after the other (performed with the primary mouse button).

As nouns, write *the right click* and *the double click* without hyphens.

Clicking while holding down a key on the keyboard

Instead of saying *hold the Ctrl key, and then click the left mouse button* you can just say *Ctrl-click*. The same applies to the Shift key and to the Alt key: *Shift-click*, *Alt-click*.

✔ **Yes:** *Ctrl-click the symbol.*

Dragging

Dragging involves clicking, and it also involves dropping the item into place. So you don't need the words *click* and *drop*.
Don't use: *put*, *move*, *drag and drop*

✘ **No:**   Click the button and drag the mouse to the toolbar.

✘ **No:**   Click the button and drag it to the toolbar.

✘ **No:**   Click and drag the button to the toolbar.

✘ **No:**   Drag and drop the button to the toolbar.

✔ **Yes:**   Drag the button to the toolbar.

Specifics with pens

If your readers might confuse the tablet pen with an ordinary pen, use *tablet pen* instead of *pen*.

Refer to the button on the side of a pen as the *pen button*.

Use *tap* and *double-tap* instead of *click* and *double-click* when describing procedures that are specific to using a pen.

Use *click* when describing a procedure that applies to using a pen as well as to using a mouse.

Specifics with touchscreens

Users *tap* with their *fingers* (not with their hands). Users can tap with one or more fingers, depending on the device and depending on the action that they're performing.

Use *tap* and *double-tap* instead of *click* and *double-click* when describing procedures that are specific to using a finger.

Use *click* when describing a procedure that applies to using a finger as well as to using a mouse.

Apply the same rules as for clicking and dragging with the mouse.

✘ **No:**   Tap and drag to view a different part of the picture.

✔ **Yes:**   Drag to view a different part of the picture.

✘ **No:**   Tap on the video that you want to play.

✔ **Yes:**   Tap the video that you want to play.

To *flick* means quickly brushing the finger to scroll through a list. Flicking is similar to dragging, but relies on momentum.

✔ **Yes:**   *Flick up or down to scroll the list.*
✔ **Yes:**   *You can browse the list with a flick.*

To *pinch* refers to the action of placing two fingers on a touch screen and then moving them closer together or farther apart. In most cases, you can just say *pinch*. If you need to describe this in more detail, you can say *pinch open* (to describe moving the fingers apart) or *pinch close* (to describe moving the fingers together).

✔ **Yes:**   *Pinch the image to zoom in or out.*
✔ **Yes:**   *Pinch open to zoom in.*

To *swipe* means to quickly slide one or more fingers across a touch screen.

✔ **Yes:**   *Swipe left over the object to delete it.*
✔ **Yes:**   *Use a three-finger swipe to page through documents.*

Related rules

Windows, messages [278]

Controls [282]

Scrolling, zooming [306]

Keys [310]

2.8.17 Scrolling, zooming

Use the following terms and phrases. Also, note the given spellings.

Scrolling

You can *scroll through* a document. Either you *scroll up*, or you *scroll down*.

You can't *scroll something*.

✘ **No:** Scroll the document to find ....

✔ **Yes:** Scroll through the document to find ....

✔ **Yes:** Scroll to view the full table of contents.

✔ **Yes:** Drag the scroll box to scroll up or down.

The whole control is the *scroll bar*.

- The part that users can drag is the *scroll box* (on Windows) or *scroller* (on Mac OS).
  Don't use: *slider*

- The *scroll arrows* are the arrow symbols at each end of the scroll bar.

Tip:
If it's clear that you mean scrolling, it's often better to use another verb such as *browse* or *move through*.

✔ **Yes:** Scroll through the list to find ....

✔ **Top:** Browse through the list to find ....

Zooming

You can:

- *zoom*

- *zoom in*

- *zoom in on* an object

- *zoom out*

Don't use: *unzoom*, *dezoom*

Related rules

Windows, messages 278

2.8.18 Resizing, aligning

Use the following terms and phrases. Also, note the given spellings.

Resizing

The *resize control* allows you to *resize* an object.

Tip:
Often, it's better to use a more general term instead of *resize*, such as *change the size of*, *enlarge*, or *make smaller*.

Aligning

Use *align* to refer to text or objects that are aligned on only one margin.

Use *justify* only to refer to text that's aligned on both the left and right margins.

Text that's *aligned on the left* is *left aligned*, text that's *aligned on the right* is *right aligned*.

You align text and objects *on a margin*, but you align them *with each other*.

Don't use: *left justified*, *right justified*

Related rules

Windows, messages 278

Controls 282

Mouse, touchscreen 302

Scrolling, zooming 306

Deleting, cutting, pasting 309

Keys 310

2.8.19 Deleting, cutting, pasting

Use the following terms and phrases. Also, note the given spellings.

Deleting / removing

Use *delete* when you mean to destroy permanently. You *delete* files from a folder, or you *delete* a paragraph from a document. (Usually, you can't restore the files, and you can't restore the deleted paragraph.)

Use *remove* when you mean *to move to another location* or *hide*. You *remove* software from the hard disk. (You probably want to reinstall later, so you don't delete the installer.) You *remove* items from a list. (You may add the item back to the list again, later.) You *remove* a toolbar that you don't need. (You may activate the toolbar again, later.)

✔ **Yes:**   Delete the file.

✔ **Yes:**   Delete the paragraph.

✔ **Yes:**   Remove the **Details** column from the **Data** window.

Don't use: *erase*, *purge*, *cut*

Cutting and pasting

Don't use *cut* as a synonym for delete.

In user documentation, avoid using *cut* or *cut and paste* as verbs or nouns. In documentation for developers and information technology professionals, however, it's OK.

✔ **Yes:**   Cut the selected text.

✔ **Yes:**   Cut and paste the selected text.

✔ **Top:**   Use the **Cut** command to delete the selected text.

✔ **Top:**   Select the text that you want to delete, and then click **Cut**.

Related rules

Windows, messages 278
Controls 282
Mouse, touchscreen 302
Scrolling, zooming 306
Keys 310

2.8.20 Keys

Use the following terms and phrases. Also, note the given spellings.

Formatting and capitalization

If your template doesn't provide any special character style for keys: Use capital letters for all key names, or add brackets around key names to indicate that you're talking about keys.

If you use brackets, use an initial capital letter if the key is usually labeled on the keyboard, but use lowercase letters if the key is usually not labeled on the keyboard.

✔ **Yes:**   *Press ALT.*
             Press [Alt].

✔ **Yes:**   *Press P.*
             Press [P].

✔ **Yes:**   *Press SPACEBAR.*
             Press [spacebar].

If your template does provide a special character style for keys: Don't add brackets. Use an initial capital letter if the key is usually labeled on the keyboard. Use lowercase letters if the key is usually not labeled on the keyboard.

✔ **Yes:**   *Press Alt.*

✔ **Yes:**   *Press P.*

✔ **Yes:**   *Press spacebar.*

Keyboard, numeric keypad

You enter text with the *keyboard*.

To enter numbers, most full-sized keyboards have a *numeric keypad* on their right side.
Don't use: *numerical keypad, numeric keyboard*

The keys on the numeric keypad are the *numeric keys*.
Don't use: *numerical keys*

On devices that don't have a keyboard, some programs show an *onscreen keyboard*.

A *soft key* is a programmable key that can invoke different functions rather than being associated with a single fixed function. Soft keys are typically located alongside a display where changing labels appear. In user documentation, it's often better to call a *soft key* just a *key*.

Pressing keys

Use *press* when pressing and immediately releasing the key initiates an action.

✔ **Yes:**   *To open help, press F1.*

Note:
When the term *press* might be confusing, you can say *use* instead of press. Example: *To move the insertion point, use the arrow keys.* (If you used *press* here, readers might think that they need to press all arrow keys simultaneously.)

Don't use *press* for onscreen elements. In this case, use *click* or *tap* (see *Mouse, touchscreen* 302).

Don't use: *depress, strike, hit, push*

Holding keys

Use *hold down* when users need to press a key until a specified action or result occurs.

✔ **Yes:**   *Hold down F8 while you restart the computer.*

Don't use *hold down* for the modifier key within a key combination.

✖ **No:**   *To open search, hold down Ctrl and press F.*

✔ **Yes:**   *To open search, press Ctrl+F.*

If you can presume that readers know that you're referring to a key on the keyboard, omit the word *key*.

✖ **No:**   *Press the F1 key.*

✔ **Yes:**   *Press F1.*

Keystroke

The act of pressing a key is a *keystroke*.
Don't use: *keypress*

Typing text, entering information

See *Controls: Fields, Boxes, Sliders* 294.

Key combinations

Indicate key combinations with a plus sign, without surrounding space characters.

Don't use a plus sign to specify a key sequence when users have to press and release each key separately. To specify a key sequence, use commas and space characters, and explain this convention.

If you have to mention only a few key combinations, and if you're writing for inexperienced users, it's often better to use a full sentence to describe the act of pressing the key combination. When writing for inexperienced users, also include the words *keys*.

Present key combinations in the following order: Windows logo key (Windows) / Command key (Mac OS) > Ctrl > Alt > Shift.

✘ No: *Press Shift+Alt+S.*
✔ Yes: *Press Alt+Shift+S.*

✔ Yes: *Press Ctrl+Alt+Del.*
✔ Yes: *Press Esc, N.*
✔ Yes: *Press the keys Ctrl, Alt, and Del simultaneously.*
✔ Yes: *Press Esc, and then immediately press N.*

Keyboard shortcuts

Keys that trigger an action or combinations of keys that trigger an action are *keyboard shortcuts*.

Don't use: *access keys, shortcut keys, accelerator keys, hot keys, speed keys, fast keys, quick keys, key combinations, key sequences*

Names of particular keys

Spell out the names of keys that could be confused or that could be difficult to read. For example, spell out: *plus sign, minus sign, hyphen, period,* and *comma*.

✘ No: *Press Shift+-.*
✔ Yes: *Press Shift+hyphen.*

Use the following key names and abbreviations:

- *spacebar, tab key* (don't use: *tab,* which is used for the tab user interface control)
- character keys: *A, ..., Z*

- modifier keys: *Alt* (don't use: *options key*), *Alt GR*, *Ctrl*, *Shift*, *Windows logo key* (Windows-specific), *Command key* (Apple-specific; don't use: *propeller key*)

- navigation keys: *Page Up*, *Page Down*, *Home*, *End*

- *arrow keys* (don't use: *direction keys*, *movement keys*): *left arrow*, *right arrow*, *up arrow*, *down arrow*

- editing keys: *Insert*, *Backspace* (on Apple keyboards: *Delete*), *Delete* (on Apple keyboards: *Forward Delete*)

- system keys: *Print Screen*, *Caps Lock* (don't use: *Shift Lock*), *Num Lock*, *Scroll Lock*

- *function keys*: *F1, ..., F12*

- other keys: *Enter*, *Return*, *Esc*, *Pause*, *Print*

Related rules

Characters 314

Windows, messages 278

Controls 282

Mouse, pen, touch screen 302

Scrolling, zooming 306

2.8.21 Characters

Use the following terms and phrases. Also, note the given spellings.

Letters in general

A letter can be a *lowercase letter* or an *uppercase letter*.
Don't use: *small letter*, *capital letter*, *capital*

Entry fields can be *case-sensitive* or *case-insensitive*.

The term *alphanumeric* refers to characters that can be either letters or numerals but not special characters, such as punctuation marks. In user documentation, it's usually better to avoid the term *alphanumeric* and to say *characters and numerals* instead.
Don't use: *alphanumerical*

Frequently used special characters

The sign ...	is called ...
( )	*parenthesis* (singular), *parentheses* (plural) *opening parenthesis*, *closing parenthesis*
{ }	*braces* *left brace, right brace* Don't use: *curly brackets*
[ ]	*brackets* *left bracket, right bracket* Don't use: *square brackets*
< >	*angle brackets* *left angle bracket, right angle bracket* Don't use: *bracket* *less-than sign* *greater-than sign*
=	*equal sign*
/	*slash*
\	*backslash*
|	*pipe* *vertical bar* *OR logical operator*
^	*caret*
~	*tilde*

#	**number sign** Use *pound key* only when referring to a telephone.
*	**asterisk**
%	**percent sign**
&	**ampersand**
@	**at sign**
'	**apostrophe**
" "	**quotation marks** *opening quotation marks, closing quotation marks* There are *straight quotation marks* and *curly quotation marks, single straight quotation marks,* and *single curly quotation marks.* For details, see *Quotation marks* 190. Don't use: *quotes, quote marks, open quotation marks, close quotation marks, beginning quotation marks, ending quotation marks*
« »	**chevrons** *opening chevron, closing chevron*
¶	**paragraph mark**
+	**plus sign**
-	**hyphen** see also *Hyphens* 183
–	**en dash** **minus sign** see also *Hyphens* 183, *Dashes* 177
—	**em dash** see also *Dashes* 177
_	**underscore**
.	**period** Don't use: *dot* (Reason: *Dot* is confusing because many fonts also include dots other than periods.)
	space character Don't use: *blank, blank character, space* (Reason: *Space* can be ambiguous because there can also be space in margins and around objects.)

Related rules

Keys 310

2.8.22   Fonts

Use the following terms and phrases. Also, note the given spellings.

Font, typeface

A *font* is a complete set of characters in one *typeface* and one *font style* (such as bold or italic).
Don't use: *face*

A *font family* consists of several fonts that share the same typeface. For example, the font family "DemoFont" might consist of the fonts "DemoFont Bold," "DemoFont Italic," "DemoFont Condensed," "DemoFont Condensed Bold," and "DemoFont Condensed Italic."
Don't use: *type family*

There are *sans-serif fonts*, *serif fonts*, and *monospaced fonts*.

Font size, font style

Typical attributes of texts are *font size* and *font style*.
Don't use: *type size*, *type style*

Possible font styles are *bold*, *italic*, and *small caps*.
Don't use: *bolded*, *boldface*, *boldfaced*, *bold type*, *italics*, *italicized*

To refer to type that's neither bold nor italic, use the term *roman*.
Don't use: *roman type*

✔ **Yes:**   *Use roman, rather than italic, for normal text.*

✔ **Yes:**   *Make the text bold.*

✔ **Yes:**   *This text is italic.*

✔ **Yes:**   *Remove the bold formatting if the characters are already bold.*

2.8.23 Printing

Use the following terms and phrases. Also, note the given spellings.

Printing

You can *print* a document.
Don't use: *print out*

You can make a *printout*. You then have a *printed document*. In addition to the *print version* of a document, there may also be an electronic version.
Don't use: *hardcopy*, *print output*

Settings

In user documentation, avoid the term *printer driver*. Use *printer software*.

You can change the *print settings* (such as the number of copies to print) and the *printer settings* (such as enabling toner-saving mode).

2.8.24   Disks, files

Use the following terms and phrases. Also, note the given spellings.

Drives, discs, disks

Common types of *drives* are:

- *hard disk drive*
- *CD drive*, *DVD drive*
- *USB drive*
- *virtual drive*
- *network drive*

Don't use *drive* when you mean *disc* or *disk*. Usually, a *drive* holds *discs* or *disks*.

Use *disc* (with the letter "c") when you refer to optical storage media, such as CDs and DVDs. Note that the word *drive* is already part of the acronym, so just say *CD* or *DVD*; don't say *CD disc* or *DVD disc*.

Use *disk* (with the letter "k") when you refer to magnetic storage media, such as *hard disks*. In user documentation, also use *disk* instead of *volume* to refer to disks and shared disks in general.

Files may also be stored on a *file server* (two words).

Don't omit the article. This is often too vague.

✗ No: *Save the file on disk.*
✗ No: *Burn the data on DVD.*
✔ Yes: *Save the file on the hard disk.*
✔ Yes: *Burn the data on a DVD.*

You *format* a disk.
Don't use: *initialize*

On the hard disk, there's free *disk space*.
Don't use: *memory*

When you mention drive letters, don't add a colon.

✗ No: *On drive C: there must be ....*
✔ Yes: *On drive C there must be ....*

Folders, files

Files are stored in *folders*. Folders may have *parent folders* and *subfolders*.

Use the terms *directory*, *parent directory*, and *subdirectory* only documentation for developers.
Don't use: *child folder*, *child directory*

A *document* is a special kind of file created and edited by the user. A *text file* is a typical document. *System files* and *program files* aren't *documents*.

A *database* (one word) is another special kind of file.

You can *convert* one file format *to* another file format.
Don't use: *convert into*

Each file and each document has a *file name*. File names usually have a *file name extension*.
Don't use: *file extension*

The *path* describes the route that the operating system follows from the *root directory* of a drive through the hierarchical structure to locate a folder or file. The path normally specifies only the drive and the hierarchical directories below the root directory. When a path also specifies a file, it's called a *full path*.
Don't use: *path name*, *home directory*

Use capital letters for abbreviations of *file types*, but use lowercase letters for *file names*.

Use capital letters for *drive names*. Capitalize the first letter of *directory names* and file names.

✔ **Yes:** *Supported formats are MP3 and MP4.*

✔ **Yes:** *They use the file name extensions .mp3 and .mp4.*

✔ **Yes:** *Open the file Music.mp3.*

✔ **Yes:** *C:\Demos\Myfile.txt*

If a file name appears at the end of a sentence, the sentence ends with a period as usual. If this could cause confusion, rewrite the sentence.

✔ **Yes:** *To view the text, open the file Readme.txt.*

✔ **Yes:** *Open the file Readme.txt to view the text.*

Don't use *Foo*, *Fu*, or *Foo.bar* as placeholders for file names. This usually doesn't make any sense to users who aren't programmers. Use a more trivial placeholder, such as *Myfile.dat*.

Open, closed, active, current

You *open* and *close* files.

Files can be *open* or *closed*.

Use *active* to refer to documents.

Use *current* to refer to drives and folders.

✔ **Yes:**   *The list shows all open files.*

✔ **Yes:**   *To save the active document, press [Ctrl]+[S].*

✔ **Yes:**   *The status bar indicates the current drive and the current folder.*

Saving data, storing data

The verbs *to save* and *to store* have slightly different connotations.

To save describes the action of putting data into the storage.

To store describes the keeping of the data.

✔ **Yes:**   *The program saves the data to the file.*

✔ **Yes:**   *The data is stored in a single file.*

✔ **Yes:**   *Store the backup copy in a safe place.*

Copying files, writing files

In user documentation, avoid the term *write* and use the term *copy* when possible. In documentation for developers and information technology professionals, the term *write* is OK.

You *copy* files *to* a disk.
Don't use: *copy on*, *copy onto*

✔ **Yes:**   In user documentation: *Copy the files to your hard disk.*

✔ **Yes:**   In documentation for developers and information technology professionals: *The application needs to write to the hard disk.*

Burning discs

You can *burn* a disc.

You can *burn a file on* a disc. (Alternatively, you can also say *burn to a disc*.)
Don't use: *burn onto a disc*

✔ **Yes:**   *You can use the program to burn CDs.*

✔ **Yes:**   *Use the program to burn your files on a CD.*

Backup

You can *back up* files (*back up* is two words).

You create a *backup copy* (*backup* is one word).

Networks

Computers are *on a network*
Don't use: *in a network*

Don't use *client* as a synonym for *computer*. Only use *client* as an adjective when referring to an object or program that obtains data from a server.

In *client/server*, use a slash, not a hyphen.

2.8.25 Internet

Use the following terms and phrases. Also, note the given spellings.

The web in general

Both spellings with and without an initial capital letter are OK: *Web* or *web*. However, be consistent. The lowercase spelling is becoming increasingly more common.

The *Internet* is always written uppercase. However, *intranet* and *extranet* are lowercase.

Don't use: *www*, *world wide web*

You can *go to the web*.

You *publish to the web* if you mean that you're generating files in a format that can be viewed online, such as HTML files.

You *publish on the web* if you mean that information or files are available online in general.

Finally, there's material *on the web* that people can read on the web or download *from the web*.

Many terms related to the web may either be written as one or as two words. However, use one spelling consistently.

- *homepage* (or *home page*)
- *web site* (or *website*)
- *web page* (or *webpage*)
- *webmaster*
- *webcam*
- *web browser* (or *browser*)
- *web server* (or *webserver*)

Online, offline

Use the term *online* (one word) only when you describe items to which the user gains access over a network. When you mean the state of being connected to a network, use a more specific term.

✘ **No:**   *When you're online ....*

✔ **Yes:**   *When you're connected to the Internet, ...*

✔ **Yes:**   *When you're logged in, ...*

✔ **Yes:** *Search the online database.*

Web sites, web pages

Don't use *web site* and *web page* interchangeably. A *web site* (such as www.example.com) usually consists of several *web pages* (such as index.html, contact.html, and so on).

You *go to a page*. You then are *at the page*. Text and images are *on the page*.

✔ **Yes:** *Go to mypage.html.*

✔ **Yes:** *You're now at the demo page.*

✔ **Yes:** *On this page, fill in the contact form, and then click **Send**.*

The *home page* is the entry page to a web site. Don't use *home page* to refer to an entire web site.

Use *visit* only to talk about going to a web site for the purpose of spending time at that site. Otherwise, use *go to*. To talk about going to a specific web page, always use *go to*.

✘ **No:** *Visit our home page.*

✔ **Yes:** *Visit our web site.*

✔ **Yes:** *Click the link to the home page.*

Browsing

You view web pages with a *browser*. Use the long form *web browser* only if you need to point out the fact that it's a browser for the web, and not another type of browser, such as an image browser.
Don't use: *Internet browser*

You *browse the Internet* or you *browse a web site*, but you *browse through* a database, list, or document.

In your browser, you can add *bookmarks* or *favorites*. In general, *bookmark* is the more common and more general term. Use *favorite* when referring to a bookmark in Internet Explorer.

URLs

In content written for end users, use the term *address* or *web address* to refer to the location of a web page. In content written for information technology professionals and software developers, use *URL*. Note that it's *a URL*, not *an URL*.

Usually, it isn't necessary to include *http://*. However, if the protocol is something other than HTTP (such as FTP), always specify the protocol with the URL.

When the URL doesn't specify a file name, omit the closing slash mark.

Use lowercase for the entire address.

URLs often appear at the end of a sentence. If there's any risk that your readers might interpret the ending period as part of the URL, rewrite the sentence or place a colon before the URL and don't include the period.

✔ **Yes:**   *Our web address is www.indoition.com.*

✔ **Yes:**   *Here you can find more information: http:// www.indoition.com/en/services/*

Links

Links take you from one web page to another web page.
Don't use: *hyperlink*, *hypertext link*, *web link*

Users *click* a link.
Don't use: *follow a link*

✖ **No:**   *Follow the download link.*

✔ **Yes:**   *Click the download link.*

Downloading, uploading

You can *download* a file from a server to your computer. The file is *downloaded*. In the opposite direction, you can *upload* a file from your hard disk to a server. The file is *uploaded*.
Don't use: *the file downloads* (a file cannot download anything), *the file uploads*

✖ **No:**   *Wait while the file downloads.*

✔ **Yes:**   *Wait while the file is being downloaded.*

✔ **Top:**   *Wait while the computer downloads the file.*

              (This is the best version because it avoids the passive and clearly states who acts.)

Email

Both spellings with and without a hyphen are OK: *email* or *e-mail*. However, use one spelling consistently. The spelling without a hyphen is becoming increasingly more common.

You *send email*.
Don't use: *to email*

Emails are stored in your *mailbox* (one word).

2.8.26 Errors

Mention limitations and problems openly, but avoid negative terminology.
Use the following terms and phrases. Also, note the given spellings.

Bug

Avoid the word *bug* because it's jargon. Instead, use *problem* or, if possible, a
more positive word such as *issue*, *condition*, or *situation*.

✘ No: This is a known bug.

✔ Yes: **This is a known problem.**

✔ Top: **This is a known issue.**

When possible, make the sentence positive. Don't put an emphasis on the
problem, but try to provide a solution instead.

✘ No: Limited availability of the update server is a known problem.

✔ Yes: **If you can't update now, try again a few minutes later.**

Use the verb *debug* only in documentation for software developers. In user
documentation, use *troubleshoot* or a more accurate word or phrase.

Failure

If a disk is *damaged*, it may *fail*.
Don't use: *corrupted*

A program may *not respond* or it may *quit unexpectedly*.
Don't use: *hang*, *crash* (crashing implies physical damage)

Valid, illegal

Don't use *illegal* when you mean *invalid*.

Use *illegal* only when you mean a violation of a law. When referring to
licenses, the terms *licensed* and *unlicensed* are more precise.

✘ No: This is an illegal file name.

✔ Yes: **The file name isn't valid.**

✔ Yes: **If the registration code isn't valid, you might have an
unlicensed copy of the program. It's illegal to use unlicensed
copies.**

Error messages

Don't put any blame on the user.

Write error messages in the third person unless you're asking for a response or are giving an instruction. This is a major difference from writing procedures, where you should avoid the passive voice and talk to the reader directly.

If it doesn't destroy confidence in your product, it's OK to put the blame on the product.

✖ No: *You can't save the file because you don't have enough free disk space. Delete some unused files.*

✔ Yes: *The file can't be saved because there isn't enough free disk space. Delete some unused files.*

✔ Top: *DemoSoft can't save the file because there isn't enough free disk space. Delete some unused files.*

(This doesn't destroy the confidence in your product because the cause of the problem isn't the product but the lack of disk space.)

✖ No: *DemoSoft has accidentally deleted all data. Please start all over again.*

(This would destroy the confidence in your product, so don't say it.)

✔ Yes: *All data has accidentally been deleted and must be re-entered.*

2.8.27 Signal words

In accordance with the specific standard that you follow (such as ANSI Z535.4 or ISO 3864-2), you must begin all warnings with a specific signal word. The signal word depends on the severity of the danger:

- A warning that begins with the signal word **CAUTION** indicates a hazard that, if not avoided, *might* result in *minor* or *moderate injury*. A warning with the signal word **CAUTION** can also refer to a situation that could damage or destroy the product or the users' work.

  If a hazard doesn't involve any risk for people but only for things, the warning symbol is often omitted.

- A warning that begins with the signal word **WARNING** indicates a hazard that, if not avoided, *could* result in *serious injury* or *death*.

- A warning that begins with the signal word **DANGER** indicates a hazard that, if not avoided, *will* result in *serious injury* or *death*.

For details, see *Writing warnings* 72.

3   References

This is the end of the book—but it's not the end of the story.

We hope that this book made you aware of the key factors that account for good user assistance, and we hope that the book will be your guide when you create your own documents.

If you want to learn more, take a look at the other books in our Technical Documentation Solutions Series:

- Technical Documentation Solutions Series: **Planning and Structuring User Assistance** — How to organize user manuals, online help systems, and other forms of user assistance in a user-friendly, easily accessible way

- Technical Documentation Solutions Series: **Designing Templates and Formatting Documents** — How to make user manuals and online help systems visually appealing and easy to read, and how to make templates efficient to use

- Technical Documentation Solutions Series: **Illustrating and Animating Help and Manuals** — How to create pictures, instruction videos, and screencasts that communicate technical information clearly

In addition, the following bibliography may be helpful. It also contains the books mentioned above. Flag symbols indicate the language of each book.

Books on technical documentation in general

Achtelig, Marc
Planning and Structuring User Assistance: How to organize user manuals, online help systems, and other forms of user assistance in a user-friendly, easily accessible way. indoition, 2012.

Achtelig, Marc
Technical Documentation Essentials: "How to Write That F***ing Manual": The essentials of technical writing in a nutshell. indoition, 2012.

Achtelig, Marc
Technische Dokumentation: „How to Write That F***ing Manual": Ohne Umschweife zu benutzerfreundlichen Handbüchern und Hilfen. Zweisprachige Ausgabe Englisch + Deutsch. indoition, 2012.

Achtelig, Marc
Translating Technical Documentation Without Losing Quality: What you shouldn't spoil when translating user manuals and online help. indoition, 2012.

Ament, Kurt
Indexing: A Nuts-and-Bolts Guide for Technical Writers. William Andrew, 2007.

Ament, Kurt
Single Sourcing: Building Modular Documentation. William Andrew, 2002.

Ballstaedt, Steffen-Peter
Wissensvermittlung. Die Gestaltung von Lernmaterial. Beltz Psychologische Verlags Union PVU, 1997.

Barker, Thomas T.
Writing Software Documentation: A Task-Oriented Approach. Longman, 2002.

Baumert, Andreas
Interviews in der Recherche: Redaktionelle Gespräche zur Informationsbeschaffung. VS Verlag für Sozialwissenschaften, 2004.

Bellamy, Laura; Carey, Michelle; Schlotfeldt, Jenifer
DITA Best Practices: A Roadmap for Writing, Editing, and Architecting in DITA. IBM, 2011.

Bellem, Birgit; Dreikorn, Johannes; Drewer, Petra; Fleury, Isabelle; Haldimann, Ralf; Jung, Martin; Keul, Udo P.; Klemm, Viktoria; Lobach, Sabine; Prusseit, Ines; Reuther, Ursula; Schmeling, Roland; Schmitz, Klaus-Dirk; Sütterlin, Volker
Leitlinie Regelbasiertes Schreiben – Deutsch für die Technische Kommunikation. tekom, 2010.

Bremer, Michael
The User Manual Manual: How to Research, Write, Test, Edit & Produce a Software Manual. Untechnical, 1999.

Brändle, Max (Hrsg.); Gabriel, Carl-Heinz (Hrsg.); Pforr, Reinhard (Hrsg); Pichler, Wolfram (Hrsg.); Schmidt, Curt (Hrsg.); Schulz, Matthias (Hrsg.)
Leitfaden für Betriebsanleitungen. tekom, 2010.

Carroll, John M. (Editor)
Minimalism Beyond the Nurnberg Funnel. The MIT Press, 1998.

Carroll, John M.
The Nurnberg Funnel: Designing Minimalist Instruction for Practical Computer Skill. The MIT Press, 1990.

Clements, Paul; Bachmann, Felix; Bass, Len; Garlan, David; Ivers, James; Little, Reed; Merson, Paulo; Nord, Robert; Stafford, Judith
Documenting Software Architectures: Views and Beyond. Addison-Wesley, 2010.

Clark, Ruth C.
Developing Technical Training: A Structured Approach for Developing Classroom and Computer-based Instructional Materials. Pfeiffer, 2007.

Clark, Ruth C.; Mayer, Richard E.
E-Learning and the Science of Instruction: Proven Guidelines for Consumers and Designers of Multimedia Learning. Pfeiffer, 2011.

Closs, Sissi
Single Source Publishing: Topicorientierte Strukturierung und DITA. entwickler press, 2006.

Coe, Marlana
Human Factors for Technical Communicators. Wiley, 1996.

Cowan, Charles
XML in Technical Communication. Institute of Scientific and Technical Communication, 2010.

DIN e.V (Herausgeber)
Technische Dokumentation: Normen für Produktdokumentation und Dokumentenmanagement. Beuth, 2008.

Drewer Petra; Ziegler, Wolfgang
Technische Dokumentation. Vogel Business Media, 2011.

Ferlein, Jörg; Hartge, Nicole
Technische Dokumentation für internationale Märkte: Haftungsrechtliche Grundlagen, Sprache, Gestaltung, Redaktion und Übersetzung. Expert, 2008.

Garrand, Timothy
Writing for Multimedia and the Web: A Practical Guide to Content Development for Interactive Media. Focal, 2006.

Gentle, Anne
Conversation and Community: The Social Web for Documentation. XML Press, 2009.

Grünwied, Gertrud
Software-Dokumentation: Grundlagen – Praxis – Lösungen. Expert, 2006.

Grupp, Josef
Handbuch Technische Dokumentation: Produktinformationen rechtskonform aufbereiten, wirtschaftlich erstellen, verständlich kommunizieren. Hanser, 2008.

Hackos, JoAnn T.
Information Development: Managing Your Documentation Projects, Portfolio, and People. Wiley, 2007.

Hackos, JoAnn T.
Introduction to DITA: A User Guide to the Darwin Information Typing Architecture. Comtech Services, 2011.

Hahn, Hans-Peter
Technische Dokumentation leichtgemacht. Hanser, 1996.

Hamilton, Richard L.
Managing Writers: A Real-World Guide to Managing Technical Documentation. XML Press, 2008.

Hargis, Gretchen; Carey, Michelle; Hernandez, Ann Kilty; Hughes, Polly; Longo, Deirdre; Rouiller, Shannon; Wilde, Elizabeth
Developing Quality Technical Information: A Handbook for Writers and Editors. IBM, 2004.

Hartman, Peter J.
Starting a Documentation Group: A Hands-On Guide. Clear Point Consultants, 1999.

Hennig, Jörg (Hrsg.), Tjarks-Sobhani, Marita (Hrsg.)
Arbeits- und Gestaltungsempfehlungen für Technische Dokumentation: Eine kritische Bestandsaufnahme. Schmidt-Römhild, 2008.

Hennig, Jörg (Hrsg.); Tjarks-Sobhani, Marita (Hrsg.)
Multimediale Technische Dokumentation. Schmidt-Römhild, 2010.

Hentrich, Johannes
DITA: Der neue Standard für Technische Dokumentation. XLcontent, 2008.

Hoffmann, Walter; Hölscher, Brigitte G.; Thiele, Ulrich
Handbuch für technische Autoren und Redakteure: Produktinformation und Dokumentation im Multimedia-Zeitalter. Publicis, 2002.

Hörmann, Hans
Meinen und Verstehen: Grundzüge einer psychologischen Semantik. Suhrkamp, 1978.

Horn, Robert E.
Mapping Hypertext: The Analysis, Organization, and Display of Knowledge for the Next Generation of On-Line Text and Graphics. Lexington, 1990.

Horton, William
Designing and Writing Online Documentation: Hypermedia for Self-Supporting Products. Wiley, 1994.

Johnston, Carol; Critcher, Ginny; Pratt, Ellis
How to write instructions. Cherryleaf, 2011.

Juhl, Dietrich
Technische Dokumentation: Anleitungen und Beispiele. Springer, 2005.

Kothes, Lars
Grundlagen der Technischen Dokumentation: Anleitungen verständlich und normgerecht erstellen. Springer, 2010.

Kühn, Cornelia
Handlungsorientierte Gestaltung von Bedienungsanleitungen. Schmidt-Römhild, 2004.

Muthig, Jürgen (Hrsg.)
Standardisierungsmethoden für die Technische Dokumentation. Schmidt-Römhild, 2008.

Pearsall, Thomas E.; Cook, Kelli Cargile
Elements of Technical Writing. Longman, 2009.

Price, Jonathan; Korman, Henry
How to Communicate Technical Information: A Handbook of Software and Hardware Documentation. Addison-Wesley Professional, 1993.

Pringle, Alan S.; O'Keefe, Sarah S.
Technical Writing 101: A Real-World Guide to Planning and Writing Technical Documentation. Scriptorium, 2009.

Rockley, Ann; Manning, Steve; Coopern Charles
Dita 101. lulu, 2009.

Rockley, Ann; Cooper, Charles
Managing Enterprise Content: A Unified Content Strategy. New Riders, 2012.

Schriver, Karen A.
Dynamics in Document Design: Creating Text for Readers. Wiley, 1996.

Schwarzman, Steven
Technical Writing Management: A Practical Guide. CreateSpace, 2011.

Self, Tony
The DITA Style Guide: Best Practices for Authors. Scriptorium, 2011.

tekom (Hrsg.)
Richtlinie zur Erstellung von Sicherheitshinweisen in Betriebsanleitungen.
tekom, 2005.

Thiele, Ulrich
Technische Dokumentationen professionell erstellen. WEKA, 2009.

Thiemann, Petra; Krings, David
Creating User-Friendly Online Help: Basics and Implementation with MadCap Flare. CreateSpace, 2009.

Tuffley, Dr. David
Software User Documentation: A How To Guide for Project Staff. CreateSpace, 2011.

Van Laan, Krista; Julian, Catherine; Hackos, JoAnn
The Complete Idiot's Guide to Technical Writing. Alpha, 2001.

Weber, Jean Hollis
Is the Help Helpful? How to Create Online Help That Meets Your Users' Needs.
Hentzenwerke, 2004.

Weber, Klaus H.
Dokumentation verfahrenstechnischer Anlagen: Praxishandbuch mit Checklisten und Beispielen. Springer, 2008

Weiß, Cornelia
Professionell dokumentieren. Beltz, 2000.

Weiss, Edmond H.
How To Write Usable User Documentation. Oryx, 1991.

Welinske, Joe
Developing User Assistance for Mobile Apps. Lulu, 2011.

Wieringa, Douglas; Barnes, Valerie E.; Moore, Christopher
Procedure Writing: Principles and Practices. Battelle, 1998.

Young, Indi
Mental Models: Aligning Design Strategy with Human Behavior. Rosenfeld, 2008.

Books on plain language and style

Achtelig, Marc
Writing Plain Instructions: How to write user manuals, online help, and other forms of user assistance that every user understands. indoition, 2012.

Achtelig, Marc
Writing Plain Instructions: Wie Sie Handbücher, Online-Hilfen und andere Formen Technischer Kommunikation schreiben, die jeder Benutzer versteht. Zweisprachige Ausgabe Englisch + Deutsch. indoition 2012.

Alred, Gerald J.; Brusaw, Charles T.; Oliu, Walter E.
Handbook of Technical Writing. St. Martin's, 2011.

Baumert, Andreas
Professionell texten: Grundlagen, Tipps und Techniken. dtv, 2008.

Blake, Gary; Bly, Robert W.
The Elements of Technical Writing. Longman, 2000.

Bremer, Michael
Untechnical Writing – How to Write About Technical Subjects and Products So Anyone Can Understand. UnTechnical, 1999.

Brogan, John A.
Clear Technical Writing. Career Education, 1973.

Burkhart, David
Stylistic traps in technical English – and solutions: Stilistische Fallen im Technischen Englisch – und Lösungen. BDÜ Fachverlag, 2010.

Bush, Donald W.
How to Edit Technical Documents. Oryx, 1995.

Goldstein, Norm
The Associated Press Stylebook and Briefing on Media Law. Basic Books, 2011.

Ikonomidis, Ageliki
Anglizismen auf gut Deutsch: Ein Leitfaden zur Verwendung von Anglizismen in deutschen Texten. Buske, 2009.

Jenkins, Jana; DeRespinis, Francis; Laird, Amy; Radzinski, Eric; McDonald, Leslie I.; Hayward, Peter
The IBM Style Guide: Conventions for Writers and Editors. Addison-Wesley Longman, 2011.

Kohl, John
The Global English Style Guide: Writing Clear, Translatable Documentation for a Global Market. SAS Press, 2008.

Langer, Inghard; Schulz von Thun, Friedemann; Tausch, Reinhard
Sich verständlich ausdrücken. Reinhardt, 2011.

Mackowiak, Klaus
Die 101 häufigsten Fehler im Deutschen: und wie man sie vermeidet. Beck, 2009.

Microsoft Corporation
Microsoft Manual of Style. Microsoft Press, 2011.

Rechenberg, Peter
Technisches Schreiben: (nicht nur) für Informatiker. Hanser, 2006.

Reiter, Markus; Sommer, Steffen
Perfekt schreiben. Hanser, 2009.

Ross-Larson, Bruce
Edit Yourself: A Manual for Everyone Who Works with Words. W. W. Norton, 1996.

Ross-Larson, Bruce
Writing for the Information Age. W. W. Norton, 2002.

Rothkegel, Annely
Technikkommunikation. UTB, 2010.

Schneider, Wolf
Deutsch für Profis: Wege zu gutem Stil. Goldmann, 2001.

Sick, Bastian
Der Dativ ist dem Genitiv sein Tod. Kiepenheuer & Witsch, 2004.

Strunk Jr., William
The Elements of Style. Longman, 1999.

Sun Technical Publications
Read Me First! A Style Guide for the Computer Industry. Prentice Hall, 2009.

University of Chicago Press
The Chicago Manual of Style. University of Chicago Press, 2010.

Weiss, Edmond H.
100 Writing Remedies: Practical Exercises for Technical Writing. Oryx, 1990.

Weiss, Edmond H.
The Elements of International English Style: A Guide to Writing Correspondence, Reports, Technical Documents, and Internet Pages for a Global Audience. M.E. Sharpe, 2005.

Weissgerber, Monika
Schreiben in technischen Berufen: Der Ratgeber für Ingenieure und Techniker: Berichte, Dokumentationen, Präsentationen, Fachartikel, Schulungsunterlagen. Publicis, 2010.

Books on graphics and design

Achtelig, Marc
Designing Templates and Formatting Documents: How to make user manuals and online help systems visually appealing and easy to read, and how to make templates efficient to use. indoition, 2012.

Alexander, Kerstin
Kompendium der visuellen Information und Kommunikation. Springer, 2007.

Ballstaedt, Steffen-Peter
Visualisieren: Über den richtigen Einsatz von Bildern. UTB, 2011.

Clark, Ruth C.; Lyons, Chopeta
Graphics for Learning: Proven Guidelines for Planning, Designing, and Evaluating Visuals in Training Materials. Pfeiffer, 2010.

Cooper, Alan
About Face: The Essentials of User Interface Design. IDG, 1999.

Cooper, Alan
The Inmates Are Running the Asylum. SAMS, 1999.

Hennig, Jörg (Herausgeber); Marita Tjarks-Sobhani (Herausgeber)
Visualisierung in Technischer Dokumentation. Schmidt-Römhild, 2004.

Horton, William
Illustrating Computer Documentation: The Art of Presenting Information Graphically on Paper and Online. Wiley, 1991.

Runk, Claudia
Grundkurs Grafik und Gestaltung. Galileo, 2010.

Williams, Robin
The Non-Designer's Type Book. Peachpit, 2005.

Williams, Robin
The Non-Designer's Design Book. Peachpit, 2008.

Wirth, Thomas
Missing Links: Über gutes Webdesign. Hanser, 2004.

4 Feedback

We sincerely hope that reading this book was a rewarding experience.

- If you like this book and think that it can help you improve your own documents, please don't hesitate to post a review and recommend the book to your colleagues. Also, don't hesitate to drop us a line. It motivates us so much to carry on :-).

- If you didn't like this book—we're embarrassed and awfully sorry. Could you please send us some feedback about what you think we should improve?

Our email address is: *feedback-WRI-1@indoition.com*

If you'd like to help us even more, please also email us your answers to the questions below.

Thank you for your support.

How to answer the questions

Please email your answers to:
feedback-WRI-1@indoition.com

For example, your email could look like this:
1c, 2a, 3b, 4a, 5d, 6a, 7a, 8b, 9b, 10a, 11a, 12b

We won't use your data for any purpose other than improving future editions of this book. If you don't want to answer all questions, that's perfectly OK. Just answer the ones that you feel comfortable with.

1. Questions about the book

How did you feel about the length of the book?

It was much too long.	1a
It was slightly too long.	1b
It was just perfect.	1c
It was slightly too short.	1d
It was much too short.	1e

Did the book cover what you'd expected, based on its title and description?

I didn't miss anything.	2a
I missed a few minor things.	2b
I missed some important points.	2c

How did you experience the depth of information?

Much of the presented information was too trivial for me.	3a
The information was just what I needed.	3b
Much of the information was too specialized for me.	3c

How did you like the practical nature of the book?

I appreciated the lack of theory and technical terms.	4a
I missed scientific background information, references to studies, and more precise terminology.	4b

Did you find any mistakes?

Yes, too many.	5a
Some, but not more than usual.	5b
Only very few.	5c
None.	5d

(If your answer is "None": Go through the book again before answering this question! No book is free from errors. If you have the time, please tell us more about the mistakes that you've found.)

2. Questions about your professional background

What's your main professional occupation?

technical writing	6a
support	6b
development	6c
marketing	6d
product management	6e
translation	6f
other	6g

How many years of experience in technical writing do you have?

less than 1	7a
1 to 3	7b
more than 3	7c

Which kind of products do you document?

mainly hardware	8a
mainly software	8b
a mixture of both hardware and software	8c

Who reads the documents that you write?

mainly consumers	9a
mainly professional users	9b

Do you speak English as a first language?

English is my first language.	10a
I speak English as a second language.	10b

Do you mainly write in English?

Yes, more than 50% of my texts are in English.	11a
No, less than 50% of my texts are in English.	11b
No, I don't write English documents at all.	11c

Who purchased this book?

I purchased the book at my own expense.	12a
The organization that I work for purchased the book.	12b
I borrowed the book from a colleague.	12c
I borrowed the book from a public library.	12d
I received a copy in a training course.	12e

Index

information types
  concept topics 46
  reference topics 49
  task topics 47
input devices
  (terms) 302
insertion point
  (terms) 302
installation
  (terms) 271
Internet
  (terms) 322
into
  vs. in 234
  vs. in to 234
intranet
  (terms) 322
invalid
  (terms) 326
isn't
  vs. is not 237
It ...
  avoiding 118
it isn't
  vs. it's not 237
it's not
  vs. it isn't 237

J

jacks
  (terms) 263
jargon
  avoiding 138
judgments
  avoiding 41

K

key combinations

(terms) 310
keyboard
  (terms) 310
keyboard shortcuts
  (terms) 310
keys
  (terms) 310
KISS principle 16

L

labels
  adding 54
large
  vs. big 206
  vs. great 206
Latin abbreviations 119
lay
  vs. lie 238
leave
  vs. let 240
LEDs
  (terms) 263
left-clicking
  (terms) 302
length
  of paragraphs 52
  of sentences 88
less
  vs. fewer 225
less than
  vs. fewer than 239
  vs. under 239
let
  vs. leave 240
letters
  (terms) 314
lie
  vs. lay 238
like

verbs
  avoiding phrasal verbs 135
  using strong verbs 132
version numbers
  handling 268
versions
  naming 268

W

want
  vs. desire 259
  vs. need 259
  vs. wish 259
warnings
  signal words 328
  writing 72
weak verbs
  avoiding 132
web
  (terms) 322
web applications
  (terms) 265
what
  vs. which 257
when
  vs. if 229
  vs. whether 229
whereas
  vs. although 258
  vs. as 258
  vs. while 258
whether
  vs. if 229
  vs. when 229
  vs. whether or not 229
whether or not
  vs. whether 229
which
  vs. that 252
  vs. what 257

while
  vs. although 258
  vs. as 258
  vs. whereas 258
who
  vs. that 254
  vs. whom 254
whom
  vs. that 254
  vs. who 254
widgets
  (terms) 265
windows
  (terms) 265, 278
wish
  vs. desire 259
  vs. need 259
  vs. want 259
within
  vs. in 235
wizards
  (terms) 276
word choice
  rules 197
word order
  general rules 96
  position of modifiers 99
words
  using common words 114
  using short words 114
  writing 113
workstations
  (terms) 263
worth
  vs. cost 216
  vs. price 216
  vs. value 216
writing
  cautions 72
  comments 76
  concept topics 46
  cross-references 81

X

Z

More Books on technical writing from indoition publishing:

Technical Documentation Basics: "How to Write That F*ing Manual"**

The essentials of technical writing in a nutshell

Planning and Structuring User Assistance

How to organize user manuals, online help systems, and other forms of user assistance in a user-friendly, easily accessible way

Designing Templates and Formatting Documents

How to make user manuals and online help systems visually appealing and easy to read, and how to make templates efficient to use

Illustrating and Animating Help and Manuals

How to create pictures, instruction videos, and screencasts that communicate technical information clearly

Translating Technical Documentation Without Losing Quality

What you shouldn't spoil when translating user manuals and online help

For detailed information, visit *www.indoition.com*.

Technical Documentation Copy and Paste Kit

Your building blocks for creating clear user assistance

The Technical Documentation Copy and Paste Kit is a substantially extended online version of this book plus all the other books of the Technical Documentation Solutions Series. The kit is your companion and style guide through **all stages of a user assistance project**:

- analysis of requirements
- structuring content
- designing templates
- writing
- illustrating
- proofreading
- translating

Like this book, the kit doesn't provide lengthy theoretical elaborations but gives practical recommendations and examples that you can easily copy and adapt to your own work.

You can install the kit on a local drive, on a network drive, or on a web server so that you and your team can access the kit **anytime, anywhere**.

You can even **add your own, company-specific comments and specifications**. To edit and manage your notes, you can use almost any HTML editor, wiki, commenting script, or content management system. You can allow all team members to edit notes and comments, or you can appoint a moderator. When you install an update of the kit, your notes and comments are fully preserved.

Effectively, you get your own, **company-specific style guide** without having all the work of setting it up and keeping it up to date.

For more information and a demo, visit *www.indoition.com*.

indoition Hotkey Script Collection for Writers and Translators

Timesaving writing macros and lookup macros that work in any program

The scripts included in the indoition Hotkey Script Collection for Writers and Translators make your work more efficient:

- **Enter frequently used words and phrases automatically** when you press a particular key combination.

- **Enter special characters** easily, such as language-specific characters and typographically correct dashes, quotation marks, and apostrophes.

- **Transform the Caps Lock key** into a regular Shift key so that when you hit Caps Lock accidentally, THINGS LIKE THIS WON'T HAPPEN ANY MORE.

- **Look up any selected term** in any online dictionary or encyclopedia with a single keystroke.

- And much more ...

You can easily edit and customize all scripts if necessary. No advanced programming skills are needed.

Unlike macros that are programmed for a specific application—such as Microsoft Word macros—the scripts in the script collection work in *all* Windows programs.

For more information and a demo, visit *www.indoition.com*.

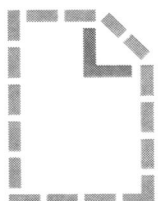

indoition Starter Template

Professional technical documentation template

Many authoring tools don't come with a suitable template for creating clear, appealing user manuals, and setting up your own template from scratch can be time-consuming. The indoition Starter Template speeds up this task and prevents you from making costly strategic mistakes. It provides:

- a design that pleases the eye *and* communicates your message clearly
- paragraph styles and character styles that are efficient to use when writing and updating your documents

The Starter Template has been designed for Microsoft Word, OpenOffice, and LibreOffice, for A4 and Letter paper sizes. If you use a different paper size, basically all you need to do is to change the page margin settings.

Many other authoring tools can import Microsoft Word files (*.docx) and OpenOffice / LibreOffice OpenDocument Text files (*.odt) as well.

Key features:

- **no bells and whistles**—the template contains only what you and your readers really need
- **automated styles** that eliminate a lot of manual formatting; optimized settings for automatic line breaks and page breaks
- **uses a time-tested**, systematic scheme for style names and keyboard shortcuts
- **works with all language versions of Microsoft Word, OpenOffice, and LibreOffice**—no need to edit style names and field codes if you're using a localized version of your authoring tool
- **well-prepared for being able to create online help from your document files as well** with the help of an appropriate single source publishing tool or converter
- includes **detailed instructions** on how to use the styles, and on how to change them if necessary

For more information, visit *www.indoition.com*.

This is *not* a happy customer:

Make them happy, write better help!

Help+Manual

www.helpandmanual.com

Help+Manual creates all standard online help formats including **HTML Help, Webhelp, PDF manuals** and **e-books** from one single source.

And it's as easy to use as a word processor.

Learn more on our website *http://www.helpandmanual.com*!

CPSIA information can be obtained
at www.ICGtesting.com
Printed in the USA
LVOW08s2133290717
543081LV00010B/108/P